T0298359

Supercontinuum Generation in Specialty Optical Fibers

This book focuses on the basic understanding of specialty optical fibers, their applications in mid-IR light generation, and cutting-edge research in the field. The book provides all the basic knowledge about specialty optical fibers and their characteristics, including dispersion, losses, propagation of modes, and so forth. Finally, the technologies based on optical fibers and their applications in all prospective areas of research are discussed.

Features:

- Provides an introduction to the history of the specialty optical fibers, and technologies based on specialty optical fibers.
- Explores specific applications of mid-IR supercontinuum generation in specialty optical fibers.
- Discusses the fabrication of specialty optical fiber-based photonic devices.
- Reviews the integration of nanotechnology with specialty optical fibers.
- Details future prospectives of specialty optical fiber-based photonic devices.

This book is aimed at graduate students and researchers in photonics, optics, physics, and photonic crystal fibers.

Supercontinuum Generation in Specialty Optical Fibers

Than Singh Saini

Ravindra Kumar Sinha

CRC Press
Taylor & Francis Group
Boca Raton London New York

CRC Press is an imprint of the
Taylor & Francis Group, an **informa** business

Designed cover image: shutterstock

First edition published 2025
by CRC Press
2385 NW Executive Center Drive, Suite 320, Boca Raton FL 33431

and by CRC Press
4 Park Square, Milton Park, Abingdon, Oxon, OX14 4RN

CRC Press is an imprint of Taylor & Francis Group, LLC

© 2025 Than Singh Saini and Ravindra Kumar Sinha

Reasonable efforts have been made to publish reliable data and information, but the author and publisher cannot assume responsibility for the validity of all materials or the consequences of their use. The authors and publishers have attempted to trace the copyright holders of all material reproduced in this publication and apologize to copyright holders if permission to publish in this form has not been obtained. If any copyright material has not been acknowledged please write and let us know so we may rectify in any future reprint.

Except as permitted under U.S. Copyright Law, no part of this book may be reprinted, reproduced, transmitted, or utilized in any form by any electronic, mechanical, or other means, now known or hereafter invented, including photocopying, microfilming, and recording, or in any information storage or retrieval system, without written permission from the publishers.

For permission to photocopy or use material electronically from this work, access www.copyright.com or contact the Copyright Clearance Center, Inc. (CCC), 222 Rosewood Drive, Danvers, MA 01923, 978-750-8400. For works that are not available on CCC please contact mpkbookspermissions@tandf.co.uk

Trademark notice: Product or corporate names may be trademarks or registered trademarks and are used only for identification and explanation without intent to infringe.

ISBN: 978-1-032-79651-2 (hbk)
ISBN: 978-1-032-81991-4 (pbk)
ISBN: 978-1-003-50240-1 (ebk)

DOI: 10.1201/9781003502401

Typeset in Times
by SPi Technologies India Pvt Ltd (Straive)

We dedicated this book to our beloved parents.
with love and regards
– T. S. Saini
– R. K. Sinha

Contents

About the Authors

Than Singh Saini obtained Ph.D. degree in nonlinear fiber optics from Delhi Technological University, Delhi, India in 2016. Currently, he is working as an Assistant Professor in the Department of Physics, National Institute of Technology, Kurukshetra, India. Prior to joining NIT Kurukshetra, he worked at national and internationally reputed institutions including CSIR-Central Scientific Instruments Organization (CSIO), Chandigarh, Optical Functional Materials Laboratory, Toyota Technological Institute, Nagoya, Japan, Indian Institute of Science Bangalore, India, University of Southampton, UK. He is working on experimental realization of specialty optical fibers for coherent mid-IR light sources, frequency comb generation, fiber optic parametric amplification, and fiber lasers. Dr. Saini is the member of SPIE and Optica (formerly OSA), USA. He has been the President of SPIE DCE Chapter for the year of 2015 and Vice President of OSA DCE Chapter at Delhi Technological University for the year of 2013–2014. He has been the recipient of the three best presentation awards for his excellent presentations in the international conferences (Awarded by SPIE and OSA). He has been awarded by the several international travel grants to present his research papers in the conferences at United States. Dr. Saini has been the academic visitor of the Edward L. Ginzton Laboratory and Stanford Nano Shared Facilities (SNSF) at Stanford University, United States in August 2014, and the College of Optical Sciences, The University of Arizona, United States in October 2014. He also attended Siegman International School on Laser at Stanford University, USA in 2014. He has also been the visitor of the International Research Centre for Nanophotonics and Metamaterials, The Metamaterials Laboratory, ITMO University (Saint-Petersburg National Research University of Information Technologies, Mechanics and Optics), Saint Petersburg-199034, Russia during March 2017. Dr. Saini is the author and co-author of more than 100 research papers published in reputed international journals and conference proceedings. Dr. Saini is the reviewer of the various international journals including *Scientific Reports, Optics Letters, Photonic Technology Letters, Applied Optics, Optik-International Journal for Light and Electron Optics, Optics Express, Optical Engineering, Optics Communications, Journal of Lightwave Technology*, and *Journal of Selected Topics in Quantum Electronics*.

 **Ravindra Kumar Sinha** completed an M.Sc. Physics from IIT Kharagpur and Ph.D. in the area of Fiber Optics and Optoelectronics from IIT Delhi. He did his Post-Doctoral Research at Osaka and Kobe university in Japan and at IISc Bangalore. He has worked at BITS Pilani, NIT Hamirpur H.P. and DCE/DTU. He has established TIFAC-CORE in Fiber Optics and Optical Communication and executed B. Tech. Engineering Physics, M. Tech.(MOCE) and M. Tech. (NST)) at DTU Delhi. Prof. Sinha has published over 390 research papers in Journals and Conference Proceedings and six book chapters and three books, filed six patents, supervised 22 sponsored projects and 21 doctoral theses. He is the Fellow of International Society of Optics and Photonics (SPIE), Fellow of IETE and Fellow of OSI. He has served as Director of CSIR-CSIO Chandigarh, CEERI Pilani and IMTECH Chandigarh. Currently, he is Vice Chancellor of Gautam Buddha University. He has mentored over 39 technology development and transfer to the industry. He is recipient of the Gold-Skoch Award for Defence Technology 2020, CSIR Technology Award 2018 and recipient of Research Excellence awards from DTU, Delhi. He is also awarded Fulbright-Nehru Fellowship 2013 as International Educational Administrator, Royal Academy of Engineering (UK) Fellowship 2008, JSPS (Japan) Fellowship and EPFL (Switzerland) Fellowship 2009, besides several awards and assignments for his research work and leadership positions in India and abroad.

Preface

The aim of the book comprises the basic to advanced level understanding of the specialty optical fibers, their applications in mid-IR light generation and the cutting-edge research up to now in the field of optical fibers. It is found that the field of non-linear fiber optics has been a most emerging area of the research for more than two decades. In the market, few books on photonic crystal fibers are already available. The photonic crystal fiber is a sort of specialty optical fibers having unique properties that cannot be achieved using conventional optical fibers. This book provides the knowledge about the various technologies in the specialty optical fibers and their applications in the field of mid-IR light generation. The book is written in an easy way to understand so that the students at all levels can understand the concepts provided in the book.

Furthermore, the book would be useful for the students who want to gain degree in fiber optics and photonics. The book covers all the basic to advanced level knowledge about the specialty optical fibers and their characteristics including dispersion, losses, propagation of modes, etc. Finally, the technologies based on optical fiber and their applications in all prospective areas of research have been discussed.

Chapter 1 provides historical development and importance of the specialty optical fibers and their applications in mid-IR supercontinuum generation. The state-of-the-art optical fibers are generally considered as optical fibers that have at least one special characteristic, distinguishing them from the standard optical fibers. One may esteem a standard optical fiber as a simple step-index optical fiber made with a standard glass material (i.e. silica) and with common values of parameters like the size of core and numerical aperture. The specialty optical fibers belong to various categories, which are discussed in details. Further, the key milestones in the development of PCF technology have been introduced. Finally, the importance of mid-IR light has been described in this chapter.

Chapter 2: In this chapter, an introduction and journey of standard optical fibers to the photonic crystal fibers have been explained. Various kinds of PCFs have been discussed. The concept of how light guide in lower reflective index core or even in hollow core fibers is provided. The fabrication techniques of different specialty optical fibers have been described in this chapter.

Chapter 3: This chapter is related to the propagation characteristics of the optical fibers that have been discussed. Principle of light guidance, modal analysis, and single mode conditions for the step-index fibers and graded index fibers have been explained. Different numerical methods and their advantages and disadvantages have been extensively described in this chapter.

Chapter 4: In this chapter, the theory of supercontinuum generation in photonic crystal fibers has been provided. The nonlinear processes involved in the broadening of supercontinuum spectrum has been discussed. Numerical modelling of supercontinuum spectrum in optical fibers has been included. Also, the description of the various physical mechanisms involved in the generation of supercontinuum light has been included in this chapter.

Preface

Chapter 5: This chapter is devoted to the various photonic crystal fiber technologies. The prospective technologies that can be evolved using the specialty optical fibers have been discussed in detail. The nanotechnologies on photonic crystal fiber platform can be implemented to obtain fiber optical devices. To increase the capacity of optical fiber links, orbital angular momentum of the light is an encouraging resource for the exploitation of the spatial dimension of the light. How the orbital angular momentum of the light can be generated using specialty optical fibers has been familiarized. Finally, the helical twisted optical fiber has been discussed in this chapter.

Chapter 6: The mid-IR supercontinuum light is expected to have numerous prospective applications including bio-photonic diagnostics, nonlinear spectroscopy, and infrared imaging. In this direction, several optical fibers and waveguide structures have been reported for broadband mid-IR supercontinuum generation. In most of the potential applications, we require intense, coherent, compact, and broadband mid-IR supercontinuum light sources. Therefore, in this chapter, the importance and applications of tapered chalcogenide optical fiber for coherent mid-IR supercontinuum generation have been discussed.

Chapter 7: As we know, the lasers are the most suitable light sources because they deliver an intense beam of coherent and collimated light. The narrowband spectral output from the laser source is greatly demanding. However, for some potential applications including medical imaging and testing of the optoelectronic devices for telecom networks, we required a broadband light source. The recent development of commercial supercontinuum sources based on optical fibers has gained much attention. Such laser sources make use of various optical nonlinear effects in specifically designed optical fibers to generate light with a wide spectrum that can span the UV-to-visible-to-mid-IR.

In this chapter, various applications of the broadband supercontinuum light generated in optical fibers are deliberated. The supercontinuum light sources are being used for various potential applications, including medical, biotechnology, military, and industry. In the medical field, the mid-IR supercontinuum light is applicable in coherence tomography, mid-IR spectroscopy, and bio-molecular sensing. The supercontinuum light sources are very suitable for detection of the cancer cells and their diagnostics. Ultrahigh-resolution optical coherence tomography using continuum generation in an air-silica photonic crystal fiber has already been demonstrated. It is also possible to show in-vivo multi-nonlinear optical imaging of a living cell using a supercontinuum light source generated from a photonic crystal fiber. For the defence sector, the mid-IR supercontinuum light offers potential applications in sensing and detecting explosive materials. Using the supercontinuum light sources, the white-light confocal microscope is conceivable for the spectrally resolved multidimensional imaging. It is possible to study photochemistry in the photonic crystal fiber-based nanoreactors using the supercontinuum light source.

Chatper 8: In this chapter, the future perspectives of the photonic crystal fibers are discussed. If somebody is interested to utilize photonics as sensing technology, then the future industry will bring a huge market for accessory instruments with numerous other benefits. There would be growing demand for additional effective cleavers, low-loss splicers, multi-port couplers, intra-fiber devices, and mode-area transformers, etc. Consequently, through PCF-based technology, we are approaching superior technology, huge job opportunity, and an improved world. As yet unexplored is the use of twisted PCF in nonlinear optics and fiber lasers, where the combination of circular and orbital angular momentum (OAM) birefringence with control of group velocity dispersion may offer opportunities for new kinds of mode-locked soliton lasers, wavelength conversion devices, and powerful supercontinuum sources. The ability of twisted fibers to provide OAM and circular birefringence suggests that yet more possibilities may emerge from this unique and unexpected guidance mechanism. The merging of nanotechnology on the PCF is exploring lab-on-fiber technology. Several other future perspectives of the PCF are also elucidated in this chapter.

We hope that this book will provide the details about the specialty optical fibers starting from its basic understanding to the advanced applications in diverse fields. Further, the book would be beneficial for the students studying in diploma, undergraduates, postgraduates courses. The scientists and engineers working in the field of specialty optical fibers and related technologies would also be benefited by this book.

Thank you.

T. S. Saini and R. K. Sinha
March 2024

1 Historical Development of Specialty Optical Fibers and Supercontinuum Generation

Specialty optical fibers (i.e. state-of-the-art optical fibers) are generally considered as optical fibers that have at least one special characteristic, distinguishing them from the *standard optical fibers*. Nevertheless, there is no completely accepted definition of the term *standard optical fiber*. One may esteem a standard optical fiber as a simple step-index optical fiber made with a standard glass material (i.e. silica) and with common values of parameters like the size of the core and numerical aperture. The specialty optical fibers belong to various categories. Some specialty optical fibers are discussed below.

- **Photonic crystal fibers (PCFs)**: The fibers where the waveguide function is not attained with the enhancement of refractive index of the core by doping rare earth elements, but in some different way. For example, the PCFs are made up by single material and the core-cladding refractive index contrast is achieved by putting the air holes in the cladding region around the core. Since the air holes are not in core region the refractive index of core region is larger than the cladding region. effective refractive index of cladding region becomes less than the core region due to air holes. Since the air holes are introduced in the cladding region that is why the PCFs are also known as the holey fibers.
- **Polarization-maintaining fibers**: The polarization-maintaining fibers can be designed and developed to maintain a linear polarization state over arbitrarily long propagation distances. In the single-polarization fibers light guides only in a single polarization direction.
- **Active fibers**: The active fibers being doped with laser-active rare earth ions are often regarded as the specialty optical fibers. In double-clad fibers with an additional larger waveguide structure for pump light, and triple-clad fibers with an additional cladding are even more special for making laser sources at various wavelengths.
- **Polyimide fibers**: Polyimide fibers are the optical fibers (with silica core and cladding) that are coated with polyimide (but usually have a glass core

DOI: 10.1201/9781003502401-1

and cladding). Such fibers can survive much higher temperatures than the conventional fibers with more common acrylate coatings.

- **Radiation-resistant fibers**: Such fibers are made up of materials that are less affected by the radiation, possibly also treated, for example, with hydrogen loading and pre-irradiation. Radiation-resistant fibers are useful in space applications and nuclear reactors.

- **Tapered optical fibers**: Tapered optical fibers can be fabricated to obtain a reduced fiber diameter along its length. Such tapered optical fibers can be used for mode field adapters and to enhance the optical nonlinear effects. Tapered optical fiber with small core acts as nonlinear optical fiber.

- **Solarization-resistant fibers**: Solarization-resistant fibers are made up of specially processed fused silica, which is relatively resistant to the ultraviolet light, circumventing excessive solarization.

- **Spun fibers**: Spun fibers are drawn from a fiber preform that is rotated around its axis during the drawing process to obtain special features in the optical fibers. They have special polarization-maintaining properties. Such optical fibers are very useful for the design and development of polarization-sensitive photonic devices.

- **Large-mode-area optical fibers**: Large-mode-area (LMA) optical fibers have a larger core size than the core size of the conventional optical fibers. The large size of the core helps to guide high power without generating optical nonlinear effects in the fiber. Therefore, LMA optical fibers are useful for the development of high-power delivery devices such as high-power fiber lasers and amplifiers.

- **Hollow-core optical fibers**: In the hollow-core optical fibers, the core is hollow instead of solid. The principle of light guidance in the hollow-core optical fibers is based on the photonic bandgap effect. Such type of optical fibers is useful for high power delivery. The nanoparticles can be deposited on the inner surface of hollow-core by inserting nanomaterial by capillary action and multifunctional photonic devices can be developed. The surface enhanced Raman scattering (SERS) in hollow-core optical fibers deposited with nanoparticles on the inner surface of core can help to detect various hazardous gases and diseases.

- **Metamaterial optical fibers**: Metamaterials are artificial photonic structures that facilitate exotic optical characteristics such as cloaking. By drawing arrays of metallic structures in a polymer fiber, it is possible to establish proper control over the electrical and magnetic resonances experienced by THz radiation transmitted through the transverse fiber cross-sections. Moreover, the use of gold in silica optical fibers for plasmonic studies can be implemented. Similarly, the soft-glasses and silicon can be implemented inside silica fibers for mid-infrared (mid-IR) nonlinear applications. The sapphire-derived all-glass optical fibers with engineered Brillouin scattering properties are very much suitable for sensors and high-power fiber laser applications.

The key milestones in the PCF technology are summarized in Table 1.1.

State-of-the-art optical fibers drawn in soft glasses (such as fluoride, tellurite, ZBLAN, and chalcogenide) are expected to have promising mediums for the design and development of high-brightness, spatially coherent mid-IR supercontinuum

TABLE 1.1
The Key Milestones in the Development of PCF Technology

S. No.	Year	Milestone/Special Characteristic	Ref.
1	1995	2D bandgaps can exist in silica/air PCF for $n_{ax} < 1$	[1]
2	1996	First solid-core PCF	[2]
3	1997	Concept of endlessly single-mode operation	[3]
4	1998	Ultra-large mode area	[4]
5	1999	Dispersion-shifted ultra-small core	[5]
6	1999	Hollow-core photonic band gap PCF	[6]
7	2000	Multi-core PCF	[7]
8	2000	Polarisation-maintaining	[8]
9	2000	Rare-earthy doped PCF laser	[9]
10	2000	Supercontinuum generation	[10]
11	2001	Carbon-dioxide laser processing of PCF	[11]
12	2001	Nondegenerate four-wave mixing	[12]
13	2001	Polymer PCF	[13]
14	2001	Soliton self-frequency shift	[14]
15	2002	Laser-tweezer guidance of particles in HC-PCF	[15]
16	2002	Long-period gratings	[16]
17	2002	PCF made from Schott SF6 glass for SC generation	[17]
18	2002	Stimulated Raman scattering in hydrogen	[18]
19	2003	Phononic bandgaps	[19]
20	2003	Rocking filters in PM PCF	[20]
21	2003	Cancellation of the soliton self-frequency shift	[21]
22	2003	Tellurite glass PCF	[22]
23	2004	All-fiber optical parametric oscillator using holey fiber	[23]
24	2004	Twin-photon generation in PCF	[24]
25	2005	EIT in acetylene	[25, 26]
26	2005	High energy transmission in HC-PCF	[27]
27	2005	Low-loss transitions between different PCFs	[28, 29]
28	2005	Photonic band gaps at 1% index contrast	[30]
29	2007	Multimaterial Multifunctional Fibers	[31]
30	2008	Index-guided (IG) PCF surface-enhanced Raman probe	[32]
31	2013	OAM optical fiber	[33]
32	2016	Helically twisted photonic crystal fibers	[34]
33	2019	Ring-core PCF for propagation of OAM modes	[35]

light sources. Supercontinuum generation (SCG) is the nonlinear process in which dramatic spectral broadening of the intense picosecond or shorter laser pulses takes place when they pass through the nonlinear condensed or gaseous optical media. Amongst non-silica glasses, the tellurite and chalcogenide glasses are considered as outstanding candidates for broadband mid-IR applications. Some of the compositions of chalcogenide glasses possess optical transparency up to 20 μm and beyond in the mid-IR region [36]. For example, As_2Se_3-based chalcogenide glass has revealed excellent optical transparency between 0.85 μm and 17.5 μm with attenuation coefficient of less than 1 cm^{-1} [36]. In addition to the broadband mid-IR transmission window, chalcogenide glasses also have very large linear and nonlinear refractive indices, which make them promising candidates for mid-IR supercontinuum applications [37]. Since last two decades, mid-IR SCG has attracted a lot of attention due to the presence of unique absorption bands of most of the bio-molecules in this domain [38]. Furthermore, mid-IR supercontinuum sources are expected to have various potential applications including mid-IR spectroscopy, bio-photonic detection and diagnostics, optical coherence tomography (OCT), multiplex coherent anti-Stokes Raman scattering microspectroscopy, and infrared imaging [39–42]. The SCG happens when numerous nonlinear optical effects including self-phase-modulation (SPM), stimulated Raman scattering (SRS), cross-phase modulation (XPM), and four-wave mixing (FWM), act together upon an intense laser pump in order to cause extensive spectral broadening of the actual pump beam.

Recently, the specialty optical fibers for mid-IR SCG and IR image transportation have been reviewed in various review articles [43–45]. Some of the substantial applications of mid-IR SCG in the spectral region from 2 μm to 20 μm are summarized below:

- **Medical applications**: In the medical field, the mid-IR SCG is useful in OCT, detection of glucose, monitoring of biomarkers in exalted breath, and the diagnostics of ulcers and colon cancer.
- **Environmental applications**: In the domain of the environmental research, the mid-IR supercontinuum light can be applicable for the understanding of the atmospheric chemistry, automobile exhaust emissions, monitoring of greenhouse gases, fire detection, and volcanic gas emissions.
- **Industrial applications**: In the industry, the mid-IR SCG can be used for the monitoring of fence lines of industrial plants and the diagnostics of the toxic and harmful gases in the semiconductor industry.
- **Mid-IR spectroscopy**: Mid-IR spectroscopic applications of supercontinuum light includes the time-resolved spectroscopy in the study of reaction kinetics, environment and climate processes, and the high-resolution spectroscopy.
- **Military and security applications**: Explosive detection, fugitive emissions from the illicit drug manufacturing sites, and sensing the toxic and harmful gases.
- **Photonic applications**: Mid-IR optical frequency metrology, characterization of the state-of-the-art specialty fiber-optics communications and infrared components based on novel fiber materials that are highly transparent in mid-IR region. Current specific inquisitiveness is the development of mid-IR light sources from 3 μm to 12 μm for LIDAR applications.

At the outset, the SCG was reported by Alfano and Shapiro in the year 1970 [46, 47]. The SCG spectrum extending from 400 nm to 700 nm was obtained in the BK7 glass sample via four-photon coupling using the picosecond laser pulses of the energy of 5 mJ at 530 nm [46]. The small-scale filaments and the frequency broadening were noticed while employing the picosecond pulse excitation in calcite, sodium chloride, quartz, and various other glasses [47]. It has been observed that the term 'supercontinuum' was not introduced in Alfano and Shapiro's work. Originally, few terms like 'superbroadening', anomalous frequency broadening, or while-light continuum' were introduced [48–51]. The term 'supercontinuum' was introduced later by the group of Manassah et al. in their works [52, 53]. It is confirmed that the white-light continuum in bulk glass is because of the formation of an optical shock at the back of the pump pulse due to self-steepening and space-time focusing [54]. The occurrence of the self-focusing is stopped at lower intensities because of the defocusing of the free-electrons and thus preventing the formation of a shock in the materials with a small band gap. In the bulk materials, the supercontinuum generation is a very complex process comprising an intricate coupling between temporal and spatial effects. On the other hand, in the case of optical fibers the process of supercontinuum generation is purely temporal dynamical processes with the transverse mode characteristics. The first experimental demonstration of supercontinuum generation in optical fiber was carried out by Lin and Stolen [55]. In the experiment, a standard silica optical fiber with zero group velocity dispersion (GVD) around 1.3 μm was used. A nanosecond dye laser is used as a pump source that offers 20 kW laser pulses. The supercontinuum spectrum with spectral bandwidth of ~200 THz on the long wavelength side of the pump was obtained. Due to the pumping in normal dispersion regime in the above experiment, the spectral broadening was dominated by the SPM. However, in the case of pumping in anomalous GVD regime, the broadening in the spectrum rises by the soliton-related dynamics. The theoretical analysis of the fibers pumped with laser pulses in anomalous GVD regime was first investigated by solving the nonlinear Schrodinger equation (NLSE) [56]. Afterward, the first experimental observation of considerable pulse compression and well-resolved pulse splitting of picosecond pulses in the negative-dispersion (normal dispersion) region of the single-mode optical fibers was reported by Mollenauer et al. [57]. This behaviour of the pulse compression and splitting was found in close agreement with the numerical calculation based on the NLSE. The picosecond pulses at 1.55 μm in optical fibers were studied for the different aspects of the higher-order soliton propagation including the pulse compression (initially 7 picoseconds to as small as 0.26 picosecond) and the unambiguous measurements of the pulse restoration over a soliton period [58, 59]. It has been observed that the optimum pulse compression was acquired for the peak-pulse input powers of several hundred watts, corresponding to moderately high (N > 10) soliton numbers. In the case of propagation of the fundamental soliton, the Raman scattering within the soliton bandwidth leads to the continuous downshift of the mean frequency known as the soliton self-frequency shift [60]. The progression of the higher-order soliton was found to be enormously perturbed by the higher-order dispersion effect, with the soliton temporally breaking up and unravelling into the fundamental soliton components. This phenomenon is generally known as 'soliton fission'. In the nonlinear optical systems, the term 'soliton fission' is an example of the symmetry breaking.

The experimental observations of the soliton fission in optical fibers revealed that it leads to an equivalent form of the supercontinuum generation in the telecommunication band [61–63]. The numerical analysis of the effect of intrapulse stimulated Raman scattering (ISRS) on the quality of the soliton effect pulse compression has been reported by solving the generalized nonlinear Schrodinger equation (GNLSE)

[64]. The numerical study showed that the ISRS can improve the performance of the soliton-effect pulse compressors both qualitatively and quantitatively. The studies showed that the soliton fission is very sensitive to the input pulse noise. As the non-linear pulse propagates in fibers its progression depends on the input power. The small fluctuations on the input pulse power lead to jitter in both the temporal and spectral characteristics of the pulse. The linking between the evolution of the higher-order soliton and the modulation instability (MI) is described by Nakazawa et al. in 1989 [65]. In this study, a perturbation theory for higher-order solitons is established to show the changes between the continuous-wave MI and the pulsed MI. Firstly, the variations in the coherence property during the progression of the supercontinuum generation in a dispersion-shifted fiber, dispersion-decreasing fiber, and a dispersion-flattened fiber have been investigated in detail [66]. The instabilities in the coherence of the SCG are due to the FWM caused by the SPM and GVD in the presence of the amplified spontaneous emission. It is well known that the incoherent light pulses are not suitable for the long-distance and high-speed transmission because it contains amplitude fluctuations and the timing jitter. Boyraz et al. demonstrate high-power, short pulse, multi-wavelength source at 10 Gb/s which was based on the spectral slic-ing of the SCG in optical fibers [67]. The coherent SCG in short fibers with the better power spectral density and uniformity, can be used for dense wavelength division multiplexing applications. However, it was also shown that the SCG in the short fibers has 13 dB higher signal-to-noise ratio (SNR) in comparison to the SCG in long fiber. Golovchenko et al. reported numerically that the supercontinuum light pro-duced by using the continuous-wave pumping in the anomalous GVD regime of the fibers also involved soliton dynamics as in the case of the pulsed excitation [68, 69]. It was revealed that the ratio of the input pump power to the power of the fundamen-tal soliton with a period of nearly the inverse of the Raman gain linewidth is of great prominence. The generation of a single soliton-like pulse is also possible from the Raman continuum. Most of the supercontinuum light sources, in general, have built by pumping with the pulsed laser systems, because very high peak power can be associated with the pulsed laser system, which helps in improving the nonlinear interactions that lead to the broadband SCG. Nevertheless, the average output power from these supercontinuum laser sources is relatively low, which results in reduced spectral density. On the contrary, the continuous-wave supercontinuum sources have shown relevant features such as impressive power spectral density, average output power, and the bandwidth with smoother spectra. The continuous-wave supercon-tinuum can be initiated by MI when an adequate amount of light is pumped at near ZDW of the medium in the anomalous dispersion region of the fiber. Resultantly, the CW light is breaking into the pulses leading to the spectral broadening, which acts as the seed for SCG. Supercontinuum light sources based on highly nonlinear fiber (HNLF) (which is a conventional silica optical fiber) offer the possibility of the splic-ing the fibers, thus forming an all-fiber architecture, allowing the high-power opera-tion. The comparison between laser beam (erbium fiber ring laser) and amplified spontaneous emission (ASE) beam-driven SCG in terms of the seed beam temporal coherence was reported by Lee et al. [70]. The pump incoherence offers a great impact on the spectral broadening of the continuum-wave SCG in the optical fibers. Under certain experimental conditions, it was demonstrated that an optimum degree

of pump incoherence produces the greatest performance in the broadening process of the continuous-wave SCG [71]. The supercontinuum source with spectral bandwidth of 1.15 to 2.15 μm was demonstrated using HNLF pumped with continuous-wave laser [72]. The numerical study of the effect of the pump coherence on the progression of continuous-wave SCG was explored by solving the GNLSE, and a design for the continuous-wave pumped SCG was recommended, linking the optimum pump bandwidth to the MI period [73]. The degree of the pump coherence has a melodramatic effect on the generating spectral broadening of supercontiuum initiated by MI in optical fibers pumped with continuous-wave pump source [74]. It has been concluded that there is a most favourable bandwidth of the pump corresponding to the optimum degree of incoherence of the pumping source, known as the function of the MI period. Chapman et al. demonstrated a continuous-wave supercontinuum source with good spectral flatness of 5 dB and average output power less than 20W in a HNLFs with small effective-mode area [75]. The higher material loss of the silica fibers and lack of the high-power laser sources for pumping at nearly 1.55 μm were the limitations on the broadband spectrum of the SCG in HNLFs. Arun et al. have demonstrated a continuous-wave SCG extending from 880 nm to 1900 nm with ~34 W output average power (with significant spectral density of >1 mW/nm from 880 nm to 1380 nm, and >50 mW/nm for 1380 nm to 1900 nm) using a standard telecom single-mode fiber (SMF-28e) with ZDW near about 1.3 μm and pumped with the high power ytterbium-doped fiber laser [76]. The performance of this high-power, continuous-wave supercontinuum light source was characterized in terms of its long-term temporal and spectral stability [77]. It was found that the continuous-wave spectrum was stable with <1 dB variation over a period of 60 minutes of uninterrupted operation. It was suggested that the small variation is ascribed to heating of the fiber used in the system and the variation can be reduced by suitably heat-dipping the fiber. After cooling the fiber down to its ambient temperature during power cycling tests, the change in spectrum was only ~0.4dB. With the purpose of pumping the telecom fiber at its ZDW (near 1.3 μm), Balaswamy et al. demonstrated a wavelength tunable grating-free, high-power, cascaded Raman laser based on a distributed feedback architecture that converts the pump power from the Ytterbium (Yb) laser operating around 1 μm into the 1.3 μm by means of a series of highly efficient cascaded Raman Stokes shifts [78]. Recently, a record output power of ~72W continuous-wave spercontinuum spectrum extending from 850 nm to 1900 nm (with a PSD of more than 3 mW/nm from 850 to 1350 nm and a very high PSD of more than 100 mW/nm from 1350 to 1900 nm) was demonstrated in a standard telecom fiber pumped by a fiber laser based on nonlinear power combining architecture [79].

In the conventional optical fibers (step-index fiber with co-axial core and cladding), the refractive index of the inner core is higher than the refractive index of the cladding. In most of the conventional optical fibers, the core-cladding index difference is about 0.1%, and various propagation features are amenable to analysis [80]. However, by establishing a microstructure in the refractive index profile of the optical fibers, there is a probability of modifying the guidance characteristics [81]. The first fabrication of a new type of optical waveguide in pure silica surrounded by a silica-air photonic crystal material with a hexagonal symmetry, known as PCF, was reported by Knight et al. in 1996 [82]. The fabricated PCF offers a single, robust, and low-loss guided mode over a broad spectral range extending from 458 nm to 1550 nm. With the success of this work, the fabrication of PCF has become technologically common-place. Depending on the geometry of the PCFs, the light can propagate by one of the two ways. In one case, if the PCF has a hollow-core at the centre of the structure genu-ine photonic band-gap guidance can happen. The photonic band-gap based PCFs have attracted abundant interest due to their prospective loss-less and distortion-free trans-mission characteristics, optical sensing, particle trapping, and innovative applications in the nonlinear fiber optics [83–86]. In the second case, if the PCF has a solid-core in the centre of the structure, the effective refractive index of the centre region of the PCF is higher than that of the surrounding photonic crystal cladding (so that the PCF contains a solid glass region surrounded by the array of air-holes running along the length of PCF), the guidance happens through the modified total internal reflation [87]. Though this is ideally analogous to the guidance mechanism in conventional optical fibers, the extra degrees of freedom attainable by altering the air-hole size and periodicity in such an index-guiding PCF. This flexibility in PCFs opens up opportu-nities to design and engineer the optical fiber waveguide properties in ways that basi-cally do not exist in standard optical fibers. The PCF with hollow-core, nevertheless, is not that used in the SCG experiments. However, gas- or liquid-filled hollow-core PCF provides more flexibility in dispersion engineering by offering additional param-eters including pressure and temperature of the gas or liquid filled in it, and therefore has been used for SCG and other applications. Most of the reports on SCG are observed in the PCFs with solid-core at the centre of the microstructure. The SCG in silica-based PCFs has been extensively reviewed by Dudley et al. in 2006 [88]. For a widespread enlightenment of SCG in PCF the readers are very much directed to go through that review article. We also encourage readers, for comprehensive under-standing of SCG mechanisms, to go through Chapter 13 of the book *Nonlinear Fiber Optics* [89]. The experimental investigation of the SCG in PCFs using low-intensity femtosecond pulses was carried out; and it was established that the reason for the white-light generation in PCFs is fission of the higher-order soliton into red-shifted fundamental solitons and blue-shifted nonsolitonic radiation [90]. The dependency of the compressibility of the supercontinuum spectrum on the input pulse-duration, spectral resolution of the pulse compressor, and the length of the PCF was examined using numerical simulations based on the stochastic extended nonlinear Schrodinger equation [91]. The supercontinuum spectrum extending from 400 nm to 2450 nm was reported in a PCF pumped with a sub-nanosecond pump source at 1064 nm [92]. Using a short length of the PCF offering two ZDWs, a 29 W continuous-wave super-continuum spectrum spanning 1.06–1.67 µm with the highest spectral power density of > 50 mW/nm upto 1.4 µm was demonstrated [93]. A continuous-wave SCG span-ning over 1300 nm with an average power upto 50 W (with the spectral power density over 50 mW/nm) was demonstrated in a PCF pumped by single-mode, continuous-wave ytterbium fiber laser of the power of 400 W at 1.07 µm [94].

The mid-IR domain of the electromagnetic spectrum is very significant because most of the molecules have their fundamental vibration absorption bands in this region. It is worthwhile to mention here that the fundamental molecular vibration bands of most of the molecules are stronger in mid-IR domain in comparison to that of in near-IR region. Mid-IR region of the spectrum extending from 2 µm to 20 µm

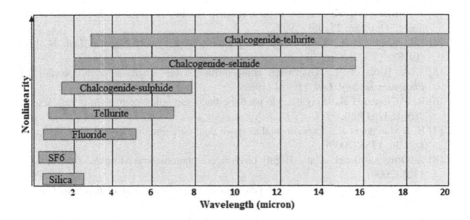

FIGURE 1.1 The transmission region of various optical fiber materials.

micron is known as molecular 'fingerprint region'. Mid-IR spectroscopy is capable to deliver thorough understanding of molecular structure of the matter and to execute non-intrusive diagnostics of composite systems of biological, physical, and chemical interest. Mid-IR light generation originates from the societal requests for environment, and health for instance, and likewise from demand for defence applications. Undeniably, the mid-IR spectral region comprises two important atmospheric transparent windows, 3–5 μm and 8–12 μm where thermal imaging (for military as well as civilian purpose) can take place. Supercontinuum spectra extending from visible to near-IR in various PCFs in silica glass have been generated for various applications. The PCFs provide very high nonlinearity and dispersion engineering to get broadband supercontinuum spectrum. However, the mid-IR range of the supercontinuum spectrum is restricted up to 2.5 μm in silica glass PCF because of its high material loss at longer wavelengths. For this reason, the development of such materials and optical fibers that offer transparency in the mid-IR range is desirable. In this context, a large number of soft-glasses (i.e. non-silica-based glasses) such as SF6, fluoride, tellurite, bismuth, ZBLAN, and chalcogenide were used for the broadband SCG in mid-IR region. The schematic of the transmission region of various optical fiber materials is illustrated in Figure 1.1.

The light guidance mechanism in solid-core PCFs is conceptually similar to that of conventional step-index optical fibers. However, in the solid-core PCF structures, extra degrees of freedom offered by modifying the air hole shape, size, orientation, and arrangement unlock the opportunities to engineer the fiber waveguide properties in ways that basically do not occur in standard step-index optical fibers.

REFERENCES

[1] T. A. Birks et al., 'Full 2D photonic band gaps in silica/air structures,' *Electron. Lett.* 31, 1941 (1995).

[2] J. C. Knight et al., *Conference on Optical Fiber Communications*, Optical Society of America (San Jose, CA), 1996.

[3] T. A. Birks, J. C. Knight, and P. St. J. Russell, 'Endlessly single-mode photonic crystal fibere,' *Opt. Lett.* 22, 961 (1997).

[4] J. C. Knight et al., 'Large mode area photonic crystal fiber,' *Electron. Lett.* 34, 1347 (1998).

[5] T. A. Birks et al., 'Dispersion compensation using single-material fibers,' *IEEE Photonics Technol. Lett.* 11, 674 (1999).

[6] R. F. Cregan et al., 'Single-mode photonic band gap guidance of light in air,' *Science* 285, 1537 (1999).

[7] B. J. Mangan et al., 'Experimental study of dual-core photonic crystal fiber,' *Electron. Lett.* 36, 1358 (2000).

[8] A. Ortigosa-Blanch et al., 'Highly birefringent photonic crystal fibers,' *Opt. Lett.* 25, 1325 (2000).

[9] W. J. Wadsworth et al., 'Yb^{3+} -doped photonic crystal fiber laser,' *Electron. Lett.* 36, 1452 (2000).

[10] J. K. Ranka, R. S. Windeler, and A. J. Stentz, 'Visible continuum generation in air-silica microstructure optical fibers with anomalous dispersion at 800 nm,' *Opt. Lett.* 25, 25 (2000).

[11] G. Kakarantzas et al., 'Miniature all-fiber devices based on CO_2 laser micro structuring of tapered fibers,' *Opt. Lett.* 26, 1137 (2001).

[12] J. E. Sharping et al., 'Four-wave mixing in microstructure fiber,' *Opt. Lett.* 26, 1048 (2001).

[13] M. A. van Eijkelenborg et al., 'Microstructured polymer optical fiber,' *Opt. Express* 9, 319 (2001).

[14] X. Liu et al., 'Soliton self-frequency shift in a short tapered air–silica microstructure fiber,' *Opt. Lett.* 26, 358 (2001).

[15] F. Benabid, J. C. Knight, and P. St. J. Russell, 'Particle levitation and guidance in hollow-core photonic crystal fiber,' *Opt. Express* 10, 1195 (2002).

[16] G. Kakarantzas, T. A. Birks, and P. St. J. Russell, Structural long-period gratings in photonic crystal fibers. *Opt. Lett.* 27, 1013 (2002).

[17] V. V. R. K. Kumar et al., 'Structural long-period gratings in photonic crystal fibers,' *Opt. Express* 10, 1520 (2002).

[18] F. Benabid et al., 'Stimulated Raman scattering in hydrogen-filled hollow-core photonic crystal Fiberfiber,' *Science* 298, 399 (2002).

[19] P. St. J. Russell et al., 'Sonic band gaps in PCF preforms: enhancing the interaction of sound and light,' *Opt. Express* 11, 2555 (2003).

[20] G. Kakarantzas et al., 'Structural rocking filters in highly birefringent photonic crystal fiber,' *Opt. Lett.* 28, 158 (2003).

[21] D. V. Skryabin et al., 'Soliton self-frequency shift cancellation in photonic crystal Fiberfibers,' *Science* 301, 1705 (2003).

[22] V. V. R. K. Kumar et al., 'Tellurite photonic crystal fiber,' *Opt. Express* 11, 2641 (2003).

[23] C. J. S. de Matos, J. R. Taylor, and K. P. Hansen, 'Continuous-wave, totally fiber integrated optical parametric oscillator using holey fiber,' *Opt. Lett.* 29, 983 (2004).

[24] J. E. Sharping et al., 'Quantum-correlated twin photons from microstructure fiber,' *Opt. Express* 12, 3086 (2004).

[25] F. Benabid et al., 'Electromagnetically-induced transparency grid in acetylene-filled hollow-core PCF,' *Opt. Express* 13, 5694 (2005).

[26] S. Ghosh et al., 'Resonant optical interactions with molecules confined in photonic band-gap Fiberfibers,' *Phys. Rev. Lett.* 94, 093902 (2005).

[27] A. H. Al-Janabi, and E. Wintner, 'High power laser transmission through photonic band gap fibers,' *Laser Phys. Lett.* 2, 137 (2005).

[28] S. G. Leon-Saval et al., 'Splice-free interfacing of photonic crystal fibers,' *Opt. Lett.* 30, 1629 (2005).

[29] A. Witkowska et al., 'All-fiber anamorphic core-shape transitions,' *Opt. Lett.* 31, 2672 (2006).

[30] A. Argyros et al., 'Photonic bandgap with an index step of one percent,' *Opt. Express* 13, 309 (2005).

[31] A. F. Abouraddy et al., 'Towards multimaterial multifunctional fibers that see, hear, sense and communicate,' *Nat. Mater.*, 6 336–347 (2007).

[32] H. Yan, J. Liu, C. Yang, G. Jin, C. Gu, and L. Hou, 'Novel index-guided photonic crystal fiber surface-enhanced Raman scattering probe,' *Opt. Express.* 16, 8300 (2008).

[33] N. Bozinovic, et al. 'Terabit-scale orbital angular momentum mode division multiplexing in fibers,' *Science* 340, 1545–1548 (2013).

[34] R. Beravat, G. K. L. Wong, M. H. Frosz, X. M. Xi and P. St. J. Russell. 'Twist-induced guidance in coreless photonic crystal fiber: A helical channel for light,' *Sci. Adv.* 2: e1601421 (2016).

[35] A. Tandje, et al., 'Ring-core photonic crystal fiber for propagation of OAM modes,' *Opt. Lett.* 44, 1611 (2019).

[36] V. Shiryaev and M. Churbanov, 'Trends and prospects for development of chalcogenide fibers for mid-infrared transmission,' *J. Non-Cryst. Solids* 377, 225–230 (2013).

[37] R. E. Slusher, G. Lenz, J. Hodelin, J. Sanghera, L. B. Shaw, and I. D. Aggarwal, 'Large Raman gain and nonlinear phase shift in high-purity As_2Se_3 chalcogenide fibers,' *J. Opt. Soc. Am. B* 21, 1146–1155 (2004).

[38] A. Schliesser, N. Picque, and T. W. Hansch, 'Mid-infrared frequency combs,' *Nat. Photon.* 6, 440–449 (2012).

[39] A. Labruyere, A. Tonello, V. Couderc, G. Huss, and P. Leproux, 'Compact supercontinuum sources and their biomedical applications,' *Opt. Fiber Technol.* 18, 375–378 (2012).

[40] R. Su, et al. 'Perspectives of mid-infrared optical coherence tomography for inspection and micrometrology of industrial ceramics,' *Opt. Express* 22, 15804–15819 (2014).

[41] F. C. Cruz, et al. 'Mid infrared optical frequency combs based on difference frequency generation for molecular spectroscopy,' *Opt. Express* 23, 26814–26824 (2015).

[42] H. N. Paulsen, K. M. Hilligse, J. Thogersen, S. R. Keiding, and J. J. Larsen, 'Coherent anti-Stokes Raman scattering microscopy with a photonic crystal fiber based light source,' *Opt. Lett.* 28(13), 1123–1125 (2003).

[43] T. S. Saini and R. K. Sinha, 'Mid-infrared supercontinuum generation in soft-glass specialty optical Fiberfibers: A review,' *Prog. Quantum Electron.* 78, 100342 (2021).

[44] T. Sylvestre, et al., 'Recent advances in supercontinuum generation in specialty optical fibers [Invited],' *J. Opt. Soc. Am.* 38, F90–F103 (2021).

[45] Y. Ohishi, 'Supercontinuum generation and IR image transportation using soft glass optical fibers: a review [Invited],' *Opt. Mater. Express* 12(10) 3990–4046 (2022).

[46] R. R. Alfano and S. L. Shapiro, 'Emission in the region 4000 to 7000 Å via four-photon coupling in glass,' *Phys. Rev. Lett.* 24, 584–587 (1970).

[47] R. R. Alfano and S. L. Shapiro, 'Observation of selfphase modulation and small-scale filaments in crystals and glasses,' *Phys. Rev. Lett.* 24, 592–594 (1970).

[48] N. G. Bondarenko, I. V. Eremina, and V. I. Talanov, 'Broadening of spectrum in self-focusing of light in crystals,' *Pis'ma Zh. Eksp. Teor. Fiz.* 12, 125–128 *JETP Lett.* 12, 85–87 (1970).

[49] N. N. Il'ichev, V. V. Korobkin, V. A. Korshunov, A. A. Malyutin, T. G. Okroashvili, and P. P. Pashinin, 'Superbroadening of the spectrum of ultrashort pulses in liquids and glasses,' *Pis'ma Zh. Eksp. Teor. Fiz.* 15, 191–194 (1972), [JETP Lett. 15, 133–135 (1972)].

[50] W. Werncke, A. Lau, M. Pfeiffer, K. Lenz, H.-J. Weigmann, and C. D. Thuy, 'An anomalous frequency broadening in water,' *Opt. Commun.* 4, 413–415 (1972).

[51] R. L. Fork, C. V. Shank, C. Hirlimann, R. Yen, and W. J. Tomlinson, 'Femtosecond white-light continuum pulses,' *Opt. Lett.* 8, 1–3 (1983).

[52] J. T. Manassah, P. P. Ho, A. Katz, and R. R. Alfano, 'Ultrafast supercontinuum laser source,' *Photon. Spectra* 18, 53–59 (1984).

[53] J. T. Manassah, R. R. Alfano, and M. Mustafa, 'Spectral distribution of an ultrafast supercontinuum laser source,' *Phys. Lett.* 107A, 305–309 (1985).

[54] A. L. Gaeta, 'Catastrophic collapse of ultrashort pulses,' *Phys. Rev. Lett.* 84, 3582–3585 (2000).

[55] C. Lin and R. H. Stolen, 'New nanosecond continuum for excited-state spectroscopy,' *Appl. Phys. Lett.* 28, 216–218 (1976).

[56] A. Hasegawa and F. Tappert, 'Transmission of stationary nonlinear optical pulses in dispersive dielectric fibers. I. Anomalous dispersion,' *Appl. Phys. Lett.* 23, 142–144 (1973).

[57] L. F. Mollenauer, R. H. Stolen, and J. P. Gordon, 'Experimental observation of picosecond pulse narrowing and solitons in optical fibers,' *Phys. Rev. Lett.* 45, 1095–1098 (1980).

[58] L. F. Mollenauer, R. H. Stolen, J. P. Gordon, and W. J. Tomlinson, 'Extreme picosecond pulse narrowing by means of soliton effect in single-mode optical fibers,' *Opt. Lett.* 8, 289–291 (1983).

[59] R. H. Stolen, L. F. Mollenauer, and W. J. Tomlinson, 'Observation of pulse restoration at the soliton period in optical fibers,' *Opt. Lett.* 8, 186–188 (1983).

[60] J. P. Gordon, 'Theory of the soliton self-frequency shift,' Opt. Lett. 11, 662–664 (1986).

[61] P. Beaud, W. Hodel, B. Zysset, and H. P. Weber, 'Ultrashort pulse propagation, pulse breakup, and fundamental soliton formation in a single-mode optical fiber,' *IEEE J. Quantum Electron.* QE-23, 1938–1946 (1987).

[62] A. S. Gouveia-Neto, M. E. Faldon, and J. R. Taylor, 'Solitons in the region of the minimum group-velocity dispersion of single-mode optical fibers,' *Opt. Lett.* 13, 770–772 (1988).

[63] M. N. Islam, G. Sucha, I. Bar-Joseph, M. Wegener, J. P. Gordon, and D. S. Chemla, 'Broad bandwidths from frequency-shifting solitons in fibers,' *Opt. Lett.* 14, 370–372 (1989).

[64] G. P. Agrawal, 'Effect of intrapulse stimulated Raman scattering on soliton-effect pulse compression in optical fibers,' *Opt. Lett.* 15, 224–226 (1990).

[65] M. Nakazawa, K. Suzuki, H. Kubota, and H. A. Haus, 'High-order solitons and the modulational instability,' *Phys. Rev. A* 39, 5768–5776 (1989).

[66] M. Nakazawa, K. R. Tamura, H. Kubota, and E. Yoshida, 'Coherence degradation in the process of supercontinuum generation in an optical fiber,' *Opt. Fiber Technol.* 4, 215–223 (1998).

[67] O. Boyraz, J. Kim, M. N. Islam, F. Coppinger, and B. Jalali, '10 Gb/ s multiple wavelength, coherent short pulse source based on spectral carving of supercontinuum generated in fibers,' *J. Lightwave Technol.* 18, 2167–2175 (2000).

[68] E. A. Golovchenko, P. V. Mamyshev, A. N. Pilipetskii, and E. M. Dianov, 'Mutual influence of the parametric effects and stimulated Raman scattering in optical fibers,' *IEEE J. Quantum Electron.* 26, 1815–1820 (1990).

[69] E. A. Golovchenko, P. V. Mamyshev, A. N. Pilipetskii, and E. M. Dianov, 'Numerical analysis of the Raman spectrum evolution and soliton pulse generation in single-mode fibers,' *J. Opt. Soc. Am. B* 8, 1626–1632 (1991).

[70] J. H. Lee, Y.-G. Han, and S. Lee, 'Experimental study on seed light source coherence dependence of continuous wave supercontinuum performance,' *Opt. Exp.* 14(8), 3443–3452 (2006).

[71] S. Martin-Lopez et al., 'Experimental investigation of the effect of pump incoherence on nonlinear pump spectral broadening and continuous-wave supercontinuum generation,' *Opt. Lett.* 31(23), 3477–3479 (2006).

[72] B. H. Chapman, S. V. Popov, and R. Taylor, 'Continuous wave supercontinuum generation through pumping in the normal dispersion region for spectral flatness,' *IEEE Photon. Technol. Lett.* 24(15), 1325–1327 (2012).

[73] E. J. R. Kelleher, 'Pump wave coherence, modulation instability and their effect on continuous-wave supercontinua,' *Opt. Fiber Technol.* 18(5), 268–282 (2012).

[74] E. J. R. Kelleher, J. C. Travers, S. V. Popov, and J. R. Taylor, 'Role of pump coherence in the evolution of continuous wave supercontinuum generation initiated by modulation instability,' *J. Opt. Soc. Amer. B, Opt. Phys.* 29, 502–512 (2012).

[75] B. H. Chapman, S. V. Popov, and R. Taylor, 'Continuous wave supercontinuum generation through pumping in the normal dispersion region for spectral flatness,' *IEEE Photonics Technol. Lett.* 24, 1325 (2012).

[76] S. Arun, V. Choudhury, V. Balaswamy, R. Prakash, and V. R. Supradeepa, 'High power, high efficiency, continuous-wave supercontinuum generation using standard telecom fibers,' *Opt. Express* 26, 7979 (2018).

[77] S. Arun, V. Choudhury, V. Balaswamy, and V. R. Supradeepa, 'Stability analysis of high power, octave spanning, continuous-wave supercontinuum sources based on cascaded Raman scattering in standard telecom fibers,' *OSA Continuum* 1, 1267 (2018).

[78] V. Balaswamy, S. Arun, S. Aparanji, V. Choudhury, and V. R. Supradeepa, 'High-power, fixed, and tunable wavelength, grating-free cascaded Raman fiber lasers,' *Opt. Lett.* 43, 1574 (2018).

[79] S. Arun, V. Choudhury, V. Balaswamy, and V. R. Supradeepa, 'Octave-spanning, continuous-wave supercontinuum generation with record power using standard telecom fibers pumped with power-combined fiber lasers,' *Opt. Lett.* 45, 1172 (2020).

[80] A. W. Snyder, and J. D. Love, Optical Waveguide Theory, Kluwer Academic (Dordrecht), 2000.

[81] P. Kaiser, and H. W. Astle, 'Low-loss single material fibers made from pure fused silica,' *Bell Syst. Tech. J.* 53, 1021–1039 (1974).

[82] J. C. Knight, T. A. Birks, P. St. J. Russell, and D. M. Atkin, 'All-silica single-mode optical fiber with photonic crystal cladding,' *Opt. Lett.* 21, 1547–1549 (1996); errata, Opt. Lett. 22, 484-485 (1997).

[83] F. Benabid, J. C. Knight, G. Antonopoulos, and P. St. J. Russell, 'Stimulated Raman scattering in hydrogen-filled hollow-core photonic crystal fiber,' *Science* 298, 399–402 (2002).

[84] J. C. Knight, 'Photonic crystal fibers,' *Nature* 424, 847–851 (2003).

[85] D. G. Ouzounov, F. R. Ahmad, D. Muller, N. Venkataraman, M. T. Gallagher, M. G. Thomas, J. Silcox, K. W. Koch, and A. L. Gaeta, 'Generation of megawatt optical solitons in hollow-core photonic band-gap fibers,' *Science* 301, 1702–1704 (2003).

[86] P. St. J. Russell, 'Photonic crystal fibers,' *Science* 299, 358–362 (2003).

[87] T. A. Birks, J. C. Knight, and P. St. J. Russell, 'Endlessly single-mode photonic crystal fiber,' *Opt. Lett.* 22, 961–963 (1997).

[88] J. M. Dudley, G. Genty, and S. Coen, 'Supercontinuum generation in photonic crystal fiber,' *Rev. Mod. Phys.* 78, 1135–1184 (2006).

[89] G. P. Agrawal, Nonlinear Fiber Optics, 5th ed., San Francisco, CA: Academic, 2013.

[90] J. Herrmann, U. Griebner, N. Zhavoronkov, A. Husakou, D. Nickel, J. C. Knight, W. J. Wadsworth, P. St. J. Russell, and G. Korn, 'Experimental evidence for supercontinuum generation by fission of higher-order solitons in photonic fibers,' *Phys. Rev. Lett.* 88(17), 173901 (2002).

[91] J. M. Dudley and S. Coen, 'Fundamental limits to few-cycle pulse generation from compression of supercontinuum spectra generated in photonic crystal fiber,' *Opt. Express* 12(11), 2423–2428 (2004).

[92] J. M. Stone and J. C. Knight, 'Visibly "white" light generation in uniform photonic crystal fiber using a microchip laser,' *Opt. Express* 16(4), 2670–2675 (2008).

[93] B. A. Cumberland, J. C. Travers, S. V. Popov, and J. R. Taylor, '29 W high power CW supercontinuum source,' *Opt. Express* 16, 5954 (2008).

[94] J. C. Travers, A. B. Rulkov, B. A. Cumberland, S. V. Popov, and J. R. Taylor, 'Visible supercontinuum generation in photonic crystal fibers with a 400 W continuous wave fiber laser,' *Opt. Exp.* 16(19), 1435–1447 (2008).

2 Introduction
Standard Optical Fibers to the Photonic Crystal Fibers

2.1 INTRODUCTION

An optical fiber is a hair-like thin rod that consists of a core of high refractive index material surrounded by a cladding region of the lower refractive index material. The transmission of information via the optical fiber over long distances with very high speed has been one of the most successful technologies of the 20th century. From the development of the first low-loss single-mode waveguides in 1970 to being vital components of the sophisticated global telecommunication network, the communication technology based on optical fiber is responsible for the high data rate transmission. The development of high-quality optical fiber technology is considered to be a leading driver behind the comparatively recent information technology revolution and the phenomenal progress in global telecommunications. Fiber optics is now engaged in all segments of telecommunication networks due to suitability for singular transmission mediums for data, voice, and video signals. Undeniably, the optical fibers have been playing a crucial role in all segments of telecommunication networks including inter-city, metro, campus, business, trans-oceanic, and transcontinental. Primary revolution in the field of fiber technology centred on realizing optical transparency in terms of manipulating the low-loss and low-dispersion transmission wavelength windows of silica single-mode optical fibers. For the first time, the low-loss optical fiber with a loss below 20 dB/km was reported in a single-mode fiber at the wavelength of He-Ne laser [1]. In the basic structure of the fiber, the high index region is surrounded by the relatively lower index uniform cladding region. Currently, the optical fibers are not only being used in long-distance communication but also have various special possible applications including high-power delivery, sensing, beam delivery for medicine and diagnostics, etc. Nevertheless, conventional step-index optical fibers have their limitations. Some of the limitations of the conventional step-index optical fiber include optical losses, optical nonlinearity, polarization effects, and group velocity dispersion. The above-mentioned drawbacks of the conventional optical fibers can be eradicated by employing newly developed special kinds of optical fibers called the specialty optical fibers or the photonic crystal fibers (PCFs). PCFs are also called holy fibers or the microstructured optical fibers). PCFs are made up of a single material in which various rings of the air holes are arranged around the core region running along the entire length. The PCF structures offer several exciting features due to the design flexibility and the high refractive index contrast between the material of PCF and air holes arranged around the core. The simplest PCF structure is illustrated in Figure 2.1. The idea of the photonic crystal is based on preventing the

DOI: 10.1201/9781003502401-2

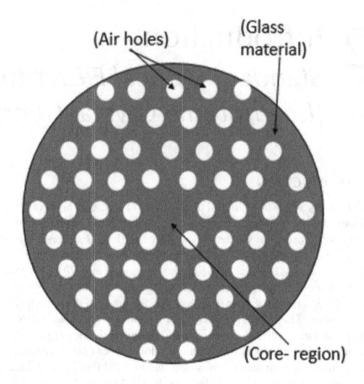

FIGURE 2.1   The schematic of the photonic crystal fiber structure.

propagation of the light of definite frequencies using periodic deviation of refractive index in the same way as the electron flow is inhibited by the periodic potential in the crystalline solids. Such inhibition of propagation in one-dimension has been used for long in the form of the periodic dielectric stacks called 'Bragg mirrors,' in which normally incident light of certain range of frequencies is not allowed to propagate within the dielectric stack. This gives rise to the photonic bandgap effect just as the same way as electronic bandgaps arises in the crystalline solids. However, the use of two-dimensional and three-dimensional periodic structures with the photonic Bandgap can be employed for the propagation in two-directions and three-directions, respectively. PCFs are two-dimensional periodic photonic structures and have been studied extensively.

During the previous three decades, lot of consideration has been made to the design and development of PCFs because of their exceptional virtues that cannot be attained using conventional step-index optical fibers [2–16]. The development of PCF was the unexpected discovery when Prof. Philip Russell and his colleagues were trying to fabricate a hollow-core optical fiber to eliminate the nonlinear optical effects in the optical fiber. In 1996, after trying various different approaches, the first silica-air solid-core PCF (single material fiber having air holes in its cladding region) was demonstrated by staking 216 silica capillaries [17]. Later, this led to the discovery of all-silica single-mode PCF within the transparency window of silica [18].

During the last decades, based on shape, size, position, arrangement, and orientation of air holes in the cladding, several PCF designs have been explored. Some of them include multi-core PCF, dispersion compensating PCF, polarization-maintaining PCF, and the rare-earth doped PCF lasers [19–21]. The schematic of the transverse cross-sectional representation of the PCF structure is shown in Figure 2.1. As shown in figure, it consists of the air holes arranged in triangular lattice pattern in the cladding region around the core. By means of the asymmetric design of the air holes and/ or the doped dielectric rods in air holes of the PCF structure, photonic scientists/ researchers are able to control the bend loss of the PCF, which makes it appropriate for the Fiber-To-The-Home technology [22].

2.2   DIFFERENT KINDS OF THE PCFs

The PCFs consist of a single material in which two-dimensional (2D) periodic structures are drawn with the variation in the plane perpendicular to the axis and invariant structure along their axis. The defect in the geometry created at the centre of the structure forms the core of the PCF. The PCFs can be divided based on the light guiding principle in the core as well as on the basis of the shape and arrangement of the air holes in the cladding region. Based on the guiding principle of light in the core region, there are two types of PCFs. One type of the PCF is the index guiding and light guides based on the principle of modified total internal reflection, while, another type of the PCF is the Bandgap PCF which is based on the photonic Bandgap guidance in the low index core or hollow-core.

2.2.1   INDEX GUIDED PCFs

The schematic of the index guiding PCF structure is illustrated in Figure 2.2. The first PCF was fabricated using pure silica material with air holes arranged in hexagonal photonic crystal cladding [17]. In the literature, since it consists of air holes, it is referred to as 'holy fiber'. The missing hole at the centre makes the core of the PCF. Therefore, such PCF can be considered to have the core of silica (refractive index, n_{co}) and the cladding with refractive index n_{cl} falling between the index of air and that of silica, depending on the relative distribution of power of the wave in

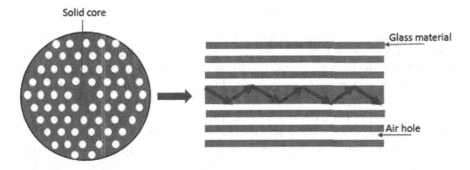

FIGURE 2.2   Index guiding photonic crystal fiber and light guiding mechanism.

the air and silica regions of the lattice. Actually, the refractive index of the cladding is the effective index of the fundamental mode supported by the infinitely extended lattice. This mode is called the 'space-filling mode (SFM)'. Since, on decreasing the wavelength, the power concentrated in the high index region increases, the cladding index increases and the effective relative index difference between core and cladding indices reduces. Therefore, the normalized frequency (V) becomes relatively insensitive to the wavelength.

Generally, in contrast to the effective refractive index of the cladding, the refractive index of the core of the conventional PCF is higher. Consequently, light is confined in the core of the PCF structure. Similar to the standard optical fiber approach, the guidance of light into the high index region of index-guided PCF is due to the modified total internal reflection (M-TIR). The total internal reflection is caused by the lower effective index in the microstructured cladding region. Since the PCF is the single material fiber, therefore, the refractive index has been lowered in the cladding region by incorporating air holes in a regular pattern so that the effective refractive index of the cladding becomes wavelength dependent. Thus, the strong wavelength dependency of the effective refractive index and the inherently large design flexibility of the PCFs allows for a whole new range of novel properties.

The mechanism of the M-TIR is equivalent to the total internal reflection and depends on a high index core region surrounded by the lower effective index provided by two-dimensional photonic crystal cladding. The effective index of such a PCF can be approximated by standard step-index fiber with a high index core and a low index cladding. Nevertheless, the effective refractive index of the cladding region of the PCFs depends on the wavelength of an optical signal propagating through these fibers. This effect permits PCFs to be designed with a whole new set of properties not conceivable using standard optical fibers. For example, the strong wavelength dependence of the refractive index of cladding allows the design of endlessly single-mode fibers, where only a single mode is supported regardless of optical wavelength. Though these fibers are named so, in practice they offer single-mode behaviour for a broad range of wavelengths covering the ultraviolet-violet-far infrared region. Moreover, it is promising to modify the dispersion characteristic of the PCFs, by this means it is plausible to design the PCFs with anomalous dispersion at visible to mid-IR wavelengths.

2.2.2   Band Gap Guided PCFs

In index-guided PCFs, the light is guided in the high index core of the fibers. It is also possible to confine the light in a core of the lower refractive index in PCFs. In this kind of PCFs, the light is generally guided in low refractive index materials by the mechanism of photonic bandgap guidance. In fact, the PCFs with an air hole at the centre have been fabricated. The first PCF with air-core at the centre was fabricated at the University of Bath, UK in 1998 [23]. The broad idea of the photonic bandgap-guided PCFs is given by Philip Russell and reported in the review article [24]. As shown in Figure 2.3, the light guides in the hollow-core or air-core by the photonic bandgap effect. The effective index of the guided mode in these fibers is less than the unity. In such PCFs, air being the core medium, the optical nonlinearities can be eliminated. Hence, very high power can be transmitted through such hollow-core

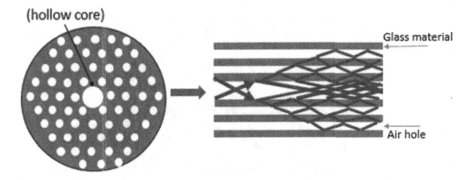

FIGURE 2.3   Photonic band gap guiding PCF and its guiding mechanism.

PCFs. Alternatively, the hollow-core can be filled with a specific gas or a liquid to have its interaction with the confined light over the long length of the PCF. This would not only be advantageous in improving the nonlinear effects in such materials but also in gas monitoring and sensing. In hollow-core PCFs, a single transverse mode is generally formed, which could have a vacuum in the core. Therefore, such PCFs offer the possibility of guiding atoms by intense light beams. The development and applications of such PCFs are discussed in the recently reported review article [7].

By the mechanism of the photonic bandgap guidance, it is possible to have light guidance in the solid core with a lower refractive index than the refractive index of the cladding. The light is confined to the lower index core; as the photonic band gap effect does not allow the propagation in the cladding region of certain frequencies. This type of approach to light guidance is different from the guiding approach using the conventional step-index optical fibers based on a well-known mechanism of total internal reflection. The schematic of the confinement of light in the low index core of the PBG fibers is illustrated in Figure 2.4.

Moreover, other types of the specialty optical fiber-based on photonic bandgap is one in which alternative layer of lower and higher refractive index materials concentrically surround a core of the lower index material or air. Such fibers are known as Bragg fibers [25]. Additionally, based on the position, size, shape, orientation, and the arrangement of the air holes in the cladding region, each of the two main types of the PCFs can be further divided into the various subclasses. For the index guiding PCFs structure, three subclasses have been immerged (i) high-numerical aperture fibers having a central part that is surrounded by a ring of relatively large air holes; (ii) large-mode-area (LMA) optical fibers consisting of relatively large dimensions and small effective refractive contrasts to spread out the transverse optical field, and (iii) highly nonlinear fibers consisting of small core dimensions to provide the tight confinement of light in the solid core [26, 27].

In the same way, the photonic bandgap optical fibers can be subdivided into two subclasses of the air-guiding, and low-index core fibers [28] or the hollow-core [29] fibers. The low-index core fibers are waveguides, which guide light by the photonic bandgap effect, and, therefore confine light to the centre of the fiber. Though the effective refractive index of the core region is lower than that of the cladding region, the major part of the optical power is propagating in high index material. In contrast

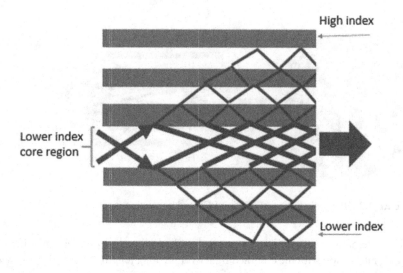

FIGURE 2.4 Schematic representation of the confinement of light in low index core of the PBG fibers.

to this, the air-guiding optical fibers provide a bandgap, which allows the mainstream of the optical power to propagate in the central hole of the fiber structure – and thereby the name air-guiding appears in such optical fibers.

2.2.3 POLARIZATION MAINTAINING PCF

As we know, light is a type of electromagnetic wave that consists of oscillating electrical fields (E) and magnetic fields (B). The properties of light can be explained by its electric and magnetic fields. As the light waves can vibrate in several directions. The light waves that are vibrating in more than one direction in more than one plane are termed as the 'unpolarized light'. The light waves that are vibrating in only one direction in a single plane are termed as 'polarized light'. In contrast to the ordinary optical fiber, where the state of polarization (SOP) varies due to bends and other deformations, polarization-maintaining (PM) fiber preserves the SOP of light. The PCFs with shorter beat lengths are better for preserving the SOP of the light. PM fibers are important in coherent optical communication and in sensing applications. In PCFs, a large index step between silica and periodic air hole helps to form birefringence, which further increases larger at the longer wavelengths. A typical PM PCF is illustrated in Figure 2.5, in which two larger size air holes are placed left and right to the core of the PCF.

2.2.4 MULTI-CORE PCFs

Multi-core PCF consists of more than one core based on the applications. To fabricate the multi-core PCFs, the process of stacking the capillaries in the region of PCF cladding makes it straightforward. The schematic representation of the transverse cross-sectional view of the multi-core PCF is shown in Figure 2.6. The seven cores in

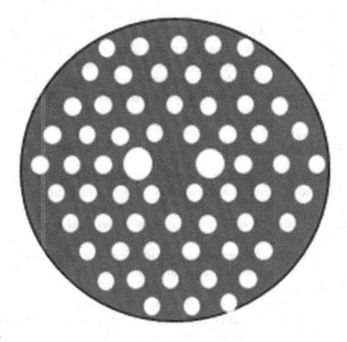

FIGURE 2.5   Schematic diagram of the polarization maintaining PCF.

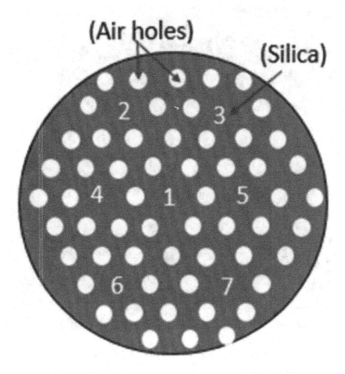

FIGURE 2.6   Multi-core photonic crystal fiber structure.

the PCF structure are represented by 1 to 7 digits in the figure. It is called multi-core due to its unique stack fabrication process. The various cores of the multi-core PCF can be designed in such a way that they permit distinct segregation of the multiple signals as well as being close enough to efficiently couple multiple signals. Each core can be placed at any position that corresponds to a lattice site, so that rectangular, linear, square, triangular, hexagonal, and other patterns are conceivable to design multi-core PCFs.

2.2.5 Highly Nonlinear PCF

Generally, an optical fiber is required to be optically linear, so that its characteristics remain unchanged no matter how much power the optical fiber carries. Nonetheless, nonlinearities in optical fibers can also be beneficial for amplifying and switching the light signal propagating through the optical fibers. At very high-power density from the laser, matters behave in a nonlinear fashion, and we come across novel phenomena including sum and difference frequency generation, second harmonic generation, four-wave-mixing, and stimulated Brillouin Scattering. The nonlinear effects can be improved by focusing light in the small core of the nonlinear PCF, as shown in Figure 2.7. Such highly nonlinear PCFs can be made to have positive dispersion at much shorter wavelengths than conventional step-index single-mode fibers [30]. Herewith, fiber solitons can be formed at the visible and near-infrared (IR) wavelengths.

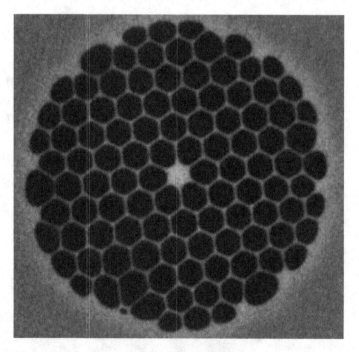

FIGURE 2.7 Highly nonlinear photonic crystal fiber structure.
Adapted from Ref. [30].

2.3   FABRICATION TECHNIQUES OF THE PCFs

The fabrication process of the PCFs is one of the most essential phases in the design and development of novel PCF-based photonic devices. The idea of producing optical fibers from a single low-loss material with microscopic air holes goes back to the early days of optical fiber technology. In 1974, Kaiser *et al.* [31] reported the first results on single material silica optical fibers. The traditional way of manufacturing optical fibers involves two main steps: fabrication of fiber-perform and drawing of the fiber using a high-temperature furnace attached in a fiber-drawing tower set-up [32]. The PCFs have been comprehended by introducing air holes in the solid glass material around the core region. In the case of the drawing process of traditional optical fibers, viscosity is an important material parameter. However, in the case of PCFs fabrication, several forces including viscosity, gravity, and surface tension are essential to be considered. Hence, the choice of the PCF material strongly influences the technological issues and the applications in the PCF fabrication process. Some methods of the fabrication of PCFs are mentioned below.

2.3.1   STACK-AND-DRAW TECHNIQUE

In the stack-and-draw technique, primarily a macroscopic scale perform is to be formed, which encompasses the structure of the interest. One possibility of creating a PCF preform is to drilling several tens to hundreds of the holes in a proper periodic arrangement into the material rod. The simple method of creating preform is the stack-and-draw method, in which the PCF preform is realized by stacking a number of capillary silica tubes and rods to form the desired air-silica structures. The stack-and-draw technique permits a high level of design flexibility because both core size and shape as well as the index profile throughout the cladding region can be controlled. After stacking of the silica capillaries and rods, the capillaries and rods are held together by thin wires and fused together during an intermediate drawing process, and the preform is drawn into preform canes. After that, the preform is drawn down on a conventional fiber-drawing tower. The temperature level of the furnace is very important to draw the PCFs since the surface tension can lead to the air hole collapse. With respect to the traditional silica optical fiber for which the usual set temperature is around 2000 to 2100 °C, a relatively lower temperature level (i.e. ~1900 °C) is enough during the drawing process of the silica PCF. A slight overpressure inside the preform is needed relative to the surroundings, in order to get careful control of the air hole size and shape during the drawing process. The time dynamics, pressure, and temperature are all very important parameters that should be precisely controlled during the PCF fabrication process. Lastly, the PCFs are coated to afford a protective standard jacket, which permits the robust handling of the PCF during its installation and uses.

2.3.2   EXTRUSION TECHNIQUE

The multi-material PCFs are fabricated using the extrusion technique. In this technique, molten glass is forced through a die containing a suitably designed pattern

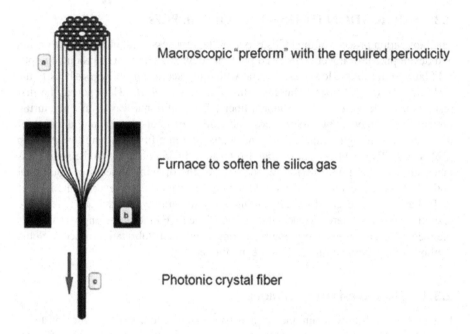

Macroscopic "preform" with the required periodicity

Furnace to soften the silica gas

Photonic crystal fiber

FIGURE 2.8   Schematic of the stack-and-draw technique of PCF fabrication.

Adapted from Ref. [7].

of holes (see Figure 2.8). The process of extrusion allows PCF to be drawn directly from the bulk glass, using a fiber-drawing tower and almost any structure, crystalline or amorphous can be produced. The extrusion is a method employed to produce objects with the fixed cross-sectional profile by pushing the soft material in a die under pressure. As illustrated in Figure 2.9, the materials in the shape of the rod, commonly known as a billet, are positioned in a sleeve held inside a furnace. Then, the billet is to be heated up to the softening temperature of the integrated materials. The appropriate pressure is supplied to push the integrated materials through the die. The pressure-assisted process imparts the shape to the extruded preform rod. The rod is subsequently drawn into the fiber form. The extrusion process works for many materials such as tellurite, chalcogenides, polymers, and compound glasses. These materials provide a lot of attractive properties, like an extended wavelength range for transmission and the superior values of the nonlinear coefficient [33].

2.3.3   THIN-FILM ROLLING TECHNIQUE

Thin-film rolling is a unique fabrication technique to incorporate the polymer materials in the preform. The process sequence of the thin-film rolling technique is shown in Figure 2.10. As illustrated in Figure 2.10, the chalcogenide glass is thermally evaporated onto both sides of the polymer film. Then, the multilayer film is rolled onto a teflon-lined mandrel, and additional polymer-cladding layers are rolled for mechanical support. The entire structure is thermally consolidated under vacuum

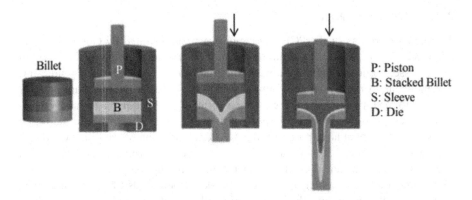

FIGURE 2.9 Schematic of the extrusion process.

Adapted form Ref. [33].

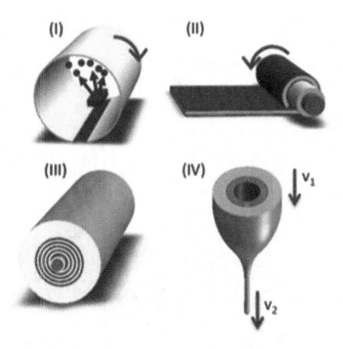

FIGURE 2.10 Schematic of the steps in the fabrication process of the photonic band gap fiber using thin-film rolling technique.

Adapted form Ref. [33].

until the materials fuse together into one solid preform. The preform is then thermally drawn into hundreds of metres of fiber by applying uniaxial tension. The ratio of the preform down-feed speed (v_1) to fiber draw speed (v_2) dictates the final layer thicknesses.

2.3.4 DIE-CAST PROCESS

The die-cast process of the PCF fabrication was demonstrated by Guiyao *et al.* in 2006 [34]. In this process of photonic crystal fiber fabrication, the die is to be put in the vessel vertically with glass material in the furnace. As shown in Figure 2.11, to remove air from the die, a vacuum hose that is connected to a vacuum pump is attached to the top of the die. The glass after softening is filled in the die under negative internal pressure. With the purpose of avoiding mixing air bubbles, the filling rate should be kept at an appropriate value. After completion of the filling of softening glass into the die, keep it for annealing to remove the thermal stress. The thermal stress can cause unwanted cracks. After reaching the die temperature close to the room temperature, the outer tube and two disks at both ends of the die are taken apart. Thereafter, to remove the heat-resisting alloy steel rods and residual debris of etched metal on the inner surface of each channel, the glass bundle is etched using an acidic solution. Finally, the etched bundle is placed in the furnace to dry for use as a PCF preform.

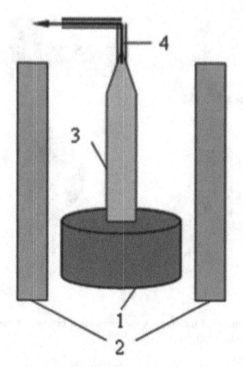

1– vessel, 2. furnace, 3. die, 4. vacuum hose.

FIGURE 2.11   Setup for the fabrication of preform to draw the PCF using die-cast process. **Adapted from Ref. [34].**

2.3.5 ROD-IN-TUBE TECHNIQUE

The soft-glass optical fibers (such as tellurite and chalcogenides glass fibers) can be fabricated using 'rod-in-tube' technique. For example, a step-index chalcogenide glass fiber with $AsSe_2$ glass as a core and As_2S_5 glass as cladding has been fabricated using the rod-in-tube technique [35]. As illustrated in Figure 2.12(a), firstly, the $AsSe_2$ glass rod and As_2S_5 glass tube can be fabricated by the casting and rotational casting method, respectively. Then the $AsSe_2$ glass rod is elongated and inserted into the As_2S_5 glass tube. In the further step, the combination of the $AsSe_2$ rod and As_2S_5 tube is elongated simultaneously in the fiber-drawing tower to obtain the step-index chalcogenide fiber. During the complete fabrication process, the nitrogen gas pressure should be adjusted as negative to avoid any interstitial hole formation between the core and the cladding. After fabrication of step-index fiber one can perform tapering using an optical fiber tapering system. The different parts of fiber including fiber cross-section, longitudinal untampered region, tapered transition region, and tapered waist regions are illustrated in Figure 2.12(b–e).

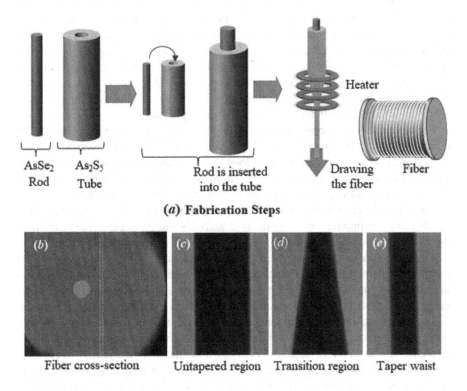

FIGURE 2.12 (a) The steps in the fabrication process of the chalcogenide step-index optical fiber using rod-in-tube method; (b) Cross-section of fabricate fiber; (c) Untapered region; (d) Transition region; (e) Taper waist.

Adapted form Ref. [35].

2.4 VARIOUS PCFs AND THEIR APPLICATIONS

As we know, PCFs have a lot of advantages over conventional optical fibers. In the PCFs, many physical properties such as power fraction, chromatic dispersion, and birefringence, etc. can be engineered. The waveguide dispersion can be engineered to have the zero-dispersion wavelength at any desired wavelength. By altering the core diameter of the fiber, the zero-dispersion wavelength can be shifted to the visible range. Also, the air holes can be filled with gases or liquids for sensing. Thereby, the tunable feature can be obtained in PCF filled with gas by changing the operating pressure of the gas. This feature can be used to design and develop very high-sensitive fiber sensors.

In practice, for compact and portable high-power devices, we need to maintain the size of the devices as small as possible. For this purpose, the LMA PCF has to be bent in the loops. Though in comparison to the conventional step-index optical fibers the bending loss of the PCF is very low, however, for LMA PCF the core size is very large ($\sim$ 40 µm or more) which increases the bend loss. This matter of high bend loss of the LMA PCFs should be carefully taken into account while fabrication. There are various techniques that are mentioned in the literature to eliminate the higher bend loss of LMA PCFs [36, 37]. Also, to escape the nonlinear effects at a high-power level, the preferred way is to use LMA PCF, which offers a dramatic reduction of the guiding intensity. The LMA PCF can be achieved by increasing the core size. However, a large core size can permit the higher-order modes to propagate through it. In a multimode fiber laser, the mode competition and intermodal dispersion can destroy the output of the fiber laser. Therefore, a fiber with LMA and single-mode operation is the preferred design for high-power applications.

For numerous other applications including optical coherence tomography (OCT), frequency comb generation, security and detection of bio-molecules white light (broadband) laser sources are needed [38–41]. As we all know that the incandescent and fluorescent lamps have a spectrum of more than 1000 nm, but, it is limited by the low intensity, lack of spatial coherence, and ordinary beam quality. Nevertheless, in the case of the laser sources that have high spatial coherence and very high brightness, which allows optimum coupling to the fiber and outstanding single-mode beam quality. However, such laser systems are limited by their monochromatic nature. If we require more than one wavelength, extra lasers of a specific wavelength are needed to cover a broadband spectrum. Using supercontinuum generation (SCG) it is expected to bridge this gap by providing ultra-broadband white-light spectrum, single-mode beam characteristics, excellent pointing stability, and brightness of a laser. The formation of the broadband spectral components can be obtained when an intense laser pulse passes through a highly nonlinear medium. The phenomenon of SCG was recounted for the first time in glass by Alfano, and Shapiro in the year 1970 [42].

The spectral domain ranging from 2 µm to 15 µm is primarily significant due to two reasons. Firstly, the important optical windows (3–5 µm and 8–13 µm) lie in which the earth's atmosphere is relatively transparent. Secondly, the strong characteristic vibration transitions of most of the molecules fall in this region [40]. The mid-infrared molecular 'fingerprint region' is significant in several essential applications in different diverse fields including medical, industry, security, and astronomy

[43–47]. Consequently, researchers are trying hard to realize mid-infrared broad-band SCG spectrum in various nonlinear conventional optical fibers, PCFs [48–51], and waveguide geometries [52–58]. It has been found that, among all non-silica glasses, the chalcogenide glasses are outstanding candidates for the application of the mid-infrared region because some of their compositions have optical transparency up to 25 μm in this domain [59]. The As_2Se_3 based chalcogenide glass has shown exceptional optical transparency within the spectral range of 0.85 – 17.5 μm with an attenuation coefficient of <1 cm^{-1} [60]. Furthermore, chalcogenide glasses have also very large linear and nonlinear refractive indices within the broadband mid-infrared transmission window, which makes them promising candidates for the mid-infrared SCG [60]. The numerical investigation of the spectral broadening of the supercontinuum spectrum extending from 2 μm to 15 μm generated in photonic crystal fiber and planar waveguide geometries has been reported [61, 62]. Also, the first experimental demonstration of the ultra-broadband supercontinuum spectrum extending from 2 μm to 15 μm in chalcogenide step-index fiber was reported in Ref. [63]. Apart from the above-mentioned applications of PCFs, they are applicable in several other applications including optical tweezers, add-drop filters for the dense wavelength division multiplexed optical communication system, generation of non-diffractive Bessel beam, fiber-based axicon devices, gas sensing, and imaging. A detailed description of various PCF structures and their applications is provided in other chapters of the book.

REFERENCES

[1] F. P. Kapron, D.B. Keck, and R.D. Maurer, 'Radiation losses in glass optical wave-guides,' App. Phys. Lett. 17, 423–425 (1970).

[2] J. C. Knight, T. A. Birks, R. F. Cregan, P. St. J. Russell and J. P. De Sandro, 'Large mode area photonic crystal fiber,' Electron. Lett. 34, 1347 (1998).

[3] M. E. Fermann, 'Single-mode excitation of multimode fibers with ultrashort pulses,' Opt. Lett. 23, 52–54 (1998).

[4] J. P. Koplow, D. A. V. Kilner and L. Goldberg, 'Single-mode operation of a coiled mul-timode fiber amplifier,' Opt. Lett. 25, 442–444 (2000).

[5] V. Rastogi and K. S. Chiang, 'Propagation characteristics of a segmented cladding fiber,' Opt. Lett. 26, 491–493 (2001).

[6] V. Rastogi and K. S. Chiang, 'Leaky optical fiber for large mode area single mode opera-tion,' Electron. Lett. 39, 1110–1112 (2003).

[7] P. Russell, 'Photonic crystal fibers,' Science 299, 358 (2003).

[8] J. Limpert, T. Schreiber, S. Nolte, H. Zellmer, A. Tunnermann, R. Iliew, F. Lederer, J. Broeng, G. Vienne, A. Petersson and C. Jakobsen, 'High-power air-clad large-mode-area photonic crystal fiber laser,' Opt. Exp. 818, 11 (2003).

[9] J. R. Folkenberg, M. D. Nielsen, N. A. Mortensen, C. Jakobsen and H. R. Simonsen, 'polarization maintaining large mode area photonic crystal fiber,' Opt. Exp. 956, 12 (2004).

[10] A. Kumar, V. Rastogi, C. Kakkar and B. Dussardier, 'Co-axial dual-core resonant leaky fiber for optical amplifiers,' J. Opt. A: Pure Appl. Opt. 10, 115306 (2008).

[11] S. Fevrier, D. Gaponov, M. Devautour, P. Roy, L. Daniault, M. Hanna, D. N. Papadopoulos, F. Druon, P. Georges, M. E. Likhachev, M. Y. Salganskii and M. V. Yashkov, 'photonic bandgap fiber oscillators and amplifiers,' Opt. Fiber Technol. 16, 419–427 (2010).

[12] B. Dussardier, V. Rastogi, A. Kumar and G. Monnom, 'Large-mode-area leaky optical fiber fabricated by MCVD,' Appl. Opt. 50, 3118–3122 (2011).

[13] J. Limpert, F. Stutzki, F. Jensen, H. J. Otto, T. Eidam, C. Jauregui and A. Tunnermann, 'Yb-doped large-pitch fibers: effective single-mode operation based on higher-order mode delocalization,' Light Sci. Appl. 1 (2012).

[14] B. L. Behera, A. Maity, S. K. Varshney and R. Datta, 'Theoretical investigations of trench-assisted large mode-area, low bend loss and single-mode microstructured core fibers,' Optics Commun. 307, 9–16 (2013).

[15] V. Rastogi and K. S. Chiang, 'Analysis of segmented-cladding fiber by the radial-effective-index method,' Opt. Soc. Am. B 21, 258–265 (2004).

[16] S. K. Varshney, K. Saitoh, M. Koshiba, B.P. Pal, and R.K. Sinha, 'Design of S-band erbium-doped, concentric dual-core photonic crystal fiber amplifiers with ASE and SRS suppression,' J. Lightwave Technol., 27, 1725–1733 (2009).

[17] J. C. Knight, T. A. Birks, D. M. Atkin, and P. St. J. Russell, 'pure silica single-mode fiber with hexagonal photonic crystal cladding,' Proc. Optical Fiber Communication 2, OSA Technical Digest Series, Optical Society of America, San Jose, CA, USA, 1996.

[18] T. A. Birks, J. C. Knight, and P. St. J. Russell, 'Endlessy single-mode photonic crystal fiber,' Opt. Lett., 22(13), 961–963 (1997).

[19] B. J. Mangan, J. C. Knight, T. A. Birks, P. St. J. Russell, and A. H. Greenaway, 'Experimental study of dual-core photonic crystal fiber,' Elect. Lett. 36, 1358–1359 (2000).

[20] A. Ortigosa-Blanch, J. C. Knight, W. J. Wadsworth, J. Arriaga, B. J. Mangan, T. A. Birks, P. St and J. Russell, 'Highly birefringent photonic crystal fibers,' Opt. Lett. 25(18), 1325–1327(2000).

[21] W. J. Wadsworth, J. C. Knight, W. H. Reeves, P. S. Russell, and J. Arriaga, 'yb^{3+}-doped photonic crystal fiber laser,' Elect. Lett. 36, 1452–1454 (2000).

[22] Y. Tsuchida, K. Saitoh, and M. Koshiba, 'design and characterization of single-mode holey fibers with low bending losses,' Opt. Exp. 13(12), 4770–4779 (2005).

[23] J. C. Knight, T.A. Birks, and P. St. J. Russell, 'Photonic bandgap guidance in optical fibers,' Science 282, 1476–1549 (1998).

[24] P. St. J. Russell and R. Dettmer, 'A neat idea [photonic crystal fiber],' IEE Rev. 47, 19–23 (2001).

[25] P. Yeh, A. Yariv, and E. Marom, 'Theory of Bragg fiber,' J. Opt. Soc. Am. 68, 1196–1201 (1978).

[26] J.C. Knight, T. A. Birks, R. F. Cregan, P. S. Russell, J. P. de Sandro, 'Large mode area photonic crystal fiber', IEE Electronics Letters 34, 1347–1348 (1998).

[27] K. P. Hansen, J. R. Jensen, C. Jacobsen, H. R. Simonsen, J. Broeng, P. M. W. Skovgaard, A. Petersson, and A. Bjarklev, 'Highly Nonlinear Photonic Crystal Fiber with Zero-Dispersion at 1.55µm,' OFC'2002, Post deadline paper, paper FA9, pp. 1–3, March, 2002.

[28] J. Broeng, S.E. Barkou, T. Sondergaard, A. Bjarklev, 'Analysis of air-guiding photonic bandgap fibers,' Optics Letters 25, 96–98 (2000).

[29] A. N. Naumov, A. M. Zheltikov, 'Optical harmonic generation in hollow-core photonic-crystal fibers: analysis of optical losses and phase matching conditions,' Quantum Electron. 32, 129–134 (2002).

[30] http://www.forc-photonics.ru/en/fibers_and_cables/Microstructured_fibers/1/160

[31] P. V. Kaiser, and H. W. Astle, 'Low-loss single-material fibers made from pure fused silica,' The Bell System Technical Journal 53, 1021–1039 (1974).

[32] A. Bjarklev, Optical fiber amplifier design and system application, Artech House (Boston, MA; London), August 1993, ISBN:0-89 006-659-0

[33] G. Tao and A. F. Abouraddy, 'Multimaterial fibers,' International Journal of Applied Glass Science 3 [4] 349–368 (2012), DOI:10.1111/ijag.12007

[34] Z. Guiyao, H. Zhiyun, L. Shuguang, and H. Lantian, 'fabrication of glass photonic crystal fibers with a die-cast process,' Appl. Opt. 45(18), 4433–4436 (2006).

[35] T. S. Saini, T. H. Tuan, T. Suzuki, Y. Ohishi, 'Coherent mid-IR supercontinuum generation using tapered chalcogenide step-index optical fiber: experiment and modelling,' Scientific Reports 10, 2236 (2020).

[36] T. S. Saini, A. Kumar, and R. K. Sinha, 'Triangular-core large-mode-area photonic crystal fiber with low bending loss for high power applications,' Applied Optics 53(31), 7246–7251 (2014).

[37] T. S. Saini, A. Kumar, and R. K. Sinha, 'Asymmetric large-mode-area photonic crystal fiber structure with effective single-mode operation: design and analysis,' Appl. Opt. 55(9), 2306–2311 (2016).

[38] J. M. Schmitt, 'Optical coherence tomography (OCT): a review,' IEEE J. Sel. Topics Quant. Elect. 5(4), 1205–1215 (1999).

[39] G. P. Agrawal, *Nonlinear Fiber Optics*, 5th ed., Elsevier Academic Press (Oxford), 2013.

[40] A. Schliesser, N. Picque, and T. W. Hansch, 'Mid-infrared frequency combs,' Nature Photon. 6, 440–449 (2012).

[41] P. Hsiung, Y. Chen, T. H. Ko, J. G. Fujimoto, C. J. S. de Matos, S. V. Popov, J. R. Taylor, and V. P. Gapontsev, 'Optical coherence tomography using a continuous-wave, high-power, Raman continuum light source,' Opt. Exp. 12, 5287–5295 (2004).

[42] R. R. Alfano, and S. L. Shapiro, 'Emission in the region 4000 to 7000 Å via four-photon coupling in glass,' Phys. Rev. Lett. 24, 584 (1970).

[43] H. Takara, T. Ohara, T. Yamamoto, H. Masuda, M. Abe, H. Takahashi, and T. Morioka, 'Field demonstration of over 1000-channel DWDM transmission with supercontinuum multi-carrier source,' Elect. Lett. 41, 270–271 (2005).

[44] S. Sanders, 'Wavelength-agile fiber laser using group-velocity dispersion of pulsed super-continua and application to broadband absorption spectroscopy,' Appl. Phys. B: Lasers and Optics 75, 799–802 (2002).

[45] A. B. Seddon, 'A prospective for new mid-infrared medical endoscopy using chalcogenide glasses,' Int. J. Appl. Glass Sci. 2, 177–191 (2011).

[46] P. Ma, D. Y. Choi, Y. Yu, X. Gai, Z. Yang, S. Debbarma, S. Madden, and B. L. Davies, 'Low-loss chalcogenide waveguides for chemical sensing in the mid-infrared,' Opt. Exp. 21(24), 29927–29937 (2013).

[47] S. Gross, N. Jovanovic, A. Sharp, M. Ireland, J. Lawrence, and M. J. Withford, 'Low loss mid-infrared ZBLAN waveguides for future astronomical applications,' Opt. Exp. 23(6), 7946–7956 (2015).

[48] N. Granzow, S. P. Stark, M. A. Schmidt, A. S. Tverjanovich, L. Wondraczek, and P. St. J. Russell, 'Supercontinuum generation in chalcogenide silica step-index fibers,' Opt. Exp. 19(21), 21003–21010 (2011).

[49] W. Gao, M. El. Amraoui, M. Liao, H. Kawashima, Z. Duan, D. Deng, T. Cheng, T. Suzuki, Y. Messaddeq, and Y. Ohishi, 'Mid-infrared supercontinuum generation in a suspended-core As_2S_3 chalcogenide microstructured optical fiber,' Opt. Exp. 21(8), 9573–9583 (2013).

[50] I. Kubat, C. S. Agger, U. Moller, A. B. Seddon, Z. Tang, S. Sujecki, T. M. Benson, David Furniss, S. Lamrini, K. Scholle, P. Fuhrberg, B. Napier, M. Farries, J. Ward, P. M. Moselund, and O. Bang, 'Mid-infrared supercontinuum generation to 12.5µm in large NA chalcogenide step-index fibers pumped at 4.5µm,' Opt. Exp. 22(16), 19169–19182 (2014).

[51] C. R. Petersen, U. Moller, I. Kubat, B. Zhou, S. Dupont, J. Ramsay, T. Besson, S. Sujecki, N. Abdel-Moneim, Z. Tang, D. Furniss, A. Seddon, and O. Bang, 'Mid-infrared supercontinuum covering the 1.4–13.3 μm molecular fingerprint region using ultra-high NA chalcogenide step-index fiber,' Nature Photon. 8, 830–834 (2014).

[52] L. Zhang, Y. Yan, Y. Yue, Q. Lin, O. Painter, R. G. Beausoleil, and A. E. Willner, 'On-chip two-octave supercontinuum generation by enhancing self-steepening of optical pulses,' Opt. Exp. 19(12), 11584–11590 (2011).

[53] R. Halir, Y. Okawachi, J. S. Levy, M. A. Foster, M. Lipson, and A. L. Gaeta, 'Ultrabroadband supercontinuum generation in a CMOS-compatible platform,' Opt. Lett. 37(10), 1685–1687 (2012).

[54] J. McCarthy, H. T. Bookey, N. D. Psaila, R. R. Thomson, and A. K. Kar, 'Mid-infrared spectral broadening in an ultrafast laser inscribed gallium lanthanum sulphide waveguide', Opt. Exp. 20, 1545–1551 (2012).

[55] J. McCarthy, H. Bookey, S. Beecher, R. Lamb, I. Elder, and A. K. Kar, 'Spectrally tailored mid-infrared super-continuum generation in a buried waveguide spanning 1750 nm to 5000 nm for atmospheric transmission,' Appl. Phys. Lett. 103, 151103 (2013).

[56] H. Hu, W. Li, N. K. Dutta, 'Dispersion-engineered tapered planar waveguide for coherent supercontinuum generation,' Opt. Commun. 324, 252–257 (2014).

[57] J. Safioui, F. Leo, B. Kuyken, S. P. Gorza, S. K. Selvaraja, R. Baets, P. Emplit, G. Roelkens, and S. Massar, 'Supercontinuum generation in hydrogenated amorphous silicon waveguides at telecommunication wavelengths,' Opt. Exp. 22(3), 3090–3097 (2014).

[58] M. R. Karim, B. M. A. Rahman, and G. P. Agrawal, 'Mid-infrared supercontinuum generation using dispersion-engineered $Ge_{11.5}As_{24}Se_{64.5}$ chalcogenide channel waveguide,' Opt. Exp. 23(5), 6903–6914 (2015).

[59] V. Shiryaev and M. Churbanov, 'Trends and prospects for development of chalcogenide fibers for mid-infrared transmission,' J. Non-Cryst. Solids 377, 225–230 (2013).

[60] R. E. Slusher, G. Lenz, J. Hodelin, J. Sanghera, L. B. Shaw, and I. D. Aggarwal, 'Large Raman gain and nonlinear phase shift in high-purity As_2Se_3 chalcogenide fibers,' J. Opt. Soc. Am. B 21, 1146–1155 (2004).

[61] T. S. Saini, A. Kumar, and R. K. Sinha, 'Broadband mid-infrared supercontinuum spectra spanning 2–15 μm using As_2Se_3 chalcogenide glass triangular-core graded-index photonic crystal fiber,' IEEE/OSA J. Lightwave Technol. 33(18), 3914–3920 (2015); DOI: 10.1109/JLT.2015.2418993;

[62] T. S. Saini, A. Kumar, and R. K. Sinha, 'Design and modeling of dispersion engineered rib waveguide for ultra broadband mid-infrared supercontinuum generation,' J. Modern Optics. 64(2), 143–149 (2017). doi:10.1080/09500340.2016.1216190;

[63] T. Cheng, K. Nagasaka, T. H. Tuan, X. Xue, M. Matsumoto, H. Tenzuka, T. Suzuki, and Y. Ohishi, 'Mid-infrared supercontinuum generation spanning 2.0 to 15.1 μm in a chalcogenide step-index fiber,' Opt. Lett. 41(9), 2117–2120 (2016).

3 Propagation Characteristics of the Optical Fibers

3.1 INTRODUCTION

As we discussed in the previous chapter, the fiber is a hair-like thin structure in which light is guided inside the core drawn along its length. Optical fiber consists of a core with a higher refractive index and cladding with a lower refractive index. The light is guided through the high refractive index core based on the principle of total internal reflection. In this chapter, the propagation characteristics, numerical methods of light propagation in optical fibers, dispersion, propagation loss, and bend loss will be discussed.

3.2 MODAL ANALYSIS

3.2.1 RAY THEORY OF LIGHT GUIDANCE

The light waves are very useful for communication. In the telecommunication system if we increase the frequency of career waves the data transmission rate can be increased. The bandwidth of the system increases. Light waves have a much larger frequency as compared to the microwave and radio waves. That is why light waves can have very large bandwidth. This is the primary reason for using light waves in high data-rate communication. If you want to use light for communication, you need to guide light across the curves and corners in the same way as copper wire conducts electricity.

3.2.2 MERIDIONAL AND SKEW RAYS

The light ray that crosses the axis of fiber is known as *meridional ray or tangential ray*. Meridional rays remain in the plane containing the optical axis and the object point from which the ray originated. The light ray that does not cross the axis of the fiber and spiral along the side of the fiber core is known as *Skew rays*. Skew ray does not parallel to the optical axis and does not lie in the plane that contains the object point and the optical axis. The skew rays never cross the axis of the optical fiber. The ray diagram of the meridional and skew rays in fiber core is illustrated in Figure 3.1.

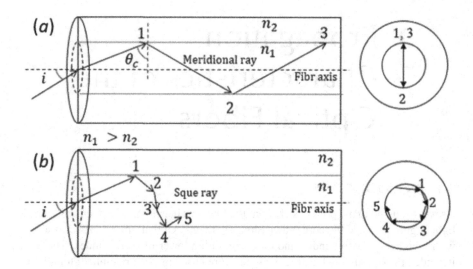

FIGURE 3.1   (a) Propagation of meridional ray, (b) propagation of sque ray.

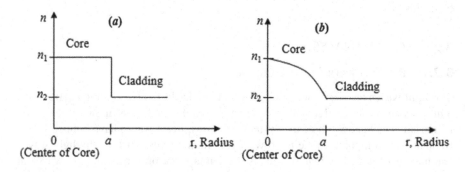

FIGURE 3.2   (a) Step-index optical fiber profile, (b) graded-index optical fiber profile.

3.2.3   STEP-INDEX AND GRADED-INDEX OPTICAL FIBERS

In the step-index optical fiber the refractive index of the core is uniform and there is a sharp decrease in the index at the cladding. The refractive index profile of step-index fiber is shown in Figure 3.2(a). However, in the case of graded-index optical fiber the refractive index profile of the core is not uniform, it decreases when we go from the centre of the core towards the outer direction. The refractive index profile of the graded-index optical fiber is shown in Figure 3.2(b). The refractive index of the core of the graded-index optical fiber varies with radius 'r'. A convenient way of defining the radially varying refractive index is by using power-law profile as given below:

$$n(r) = n_1 \sqrt{\left[1 - 2\Delta \left(\frac{r}{a}\right)^\alpha\right]} \quad \text{for } 0 \leq r \leq a$$

$$= n_1 \sqrt{[1 - 2\Delta]} \approx n_1 [1 - \Delta] = n_2 \quad \text{for } r \geq a \tag{3.1}$$

where '*a*' is the radius of the core, and *r* represents the radial distance for the axis of the core.

3.2.4 STRUCTURE OF THE STEP-INDEX OPTICAL FIBER

In the following Figure 3.3(a) the structure of step-index optical fiber is shown. It consists of a cylindrical core with high refractive index of n_1 with radius '*a*', and around the core there is cladding with a relatively lower refractive index of n_2 with radius '*b*'. The refractive index profile of the core and cladding is uniform and shown in Figure 3.3(b).

Core: 0 <*r*>*a*; cladding: a <*r*>*b*; *a* is the core radius, and *b*-represents the radius of the cladding.

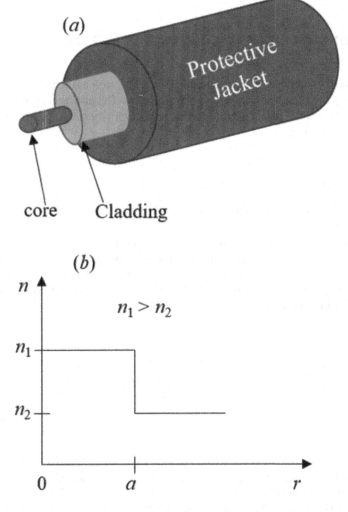

FIGURE 3.3   (a) Structure of an optical fiber, (b) refractive index profile of an optical fiber.

A parameter Δ through the following equations
Refractive index difference-

$$\Delta = \frac{n_1^2 - n_2^2}{2n_1^2} \tag{3.2}$$

For weekly guidance fiber ($\Delta \ll 1$) (it is true for silica fibers where n_1 is very nearly equal to n_2) we may write $n_1 \approx n_2$

$$\Delta \approx \frac{n_1^2 - n_2^2}{2n_1^2} \approx \frac{(n_1 + n_2)(n_1 - n_2)}{2n_1^2} \approx \frac{n_1 - n_2}{n_1} \approx \frac{n_1 - n_2}{n_2} \tag{3.3}$$

The index level of the doped glass is frequently characterized by Δ, and given in percentage.

3.3 PRINCIPLE OF LIGHT GUIDANCE IN STEP-INDEX OPTICAL FIBER

In the optical communication system, the optical fiber acts as the transmission channel carrying the light beam loaded with information. The principle by which light is guided in the fiber is **Total Internal Reflection** (TIR). You know the refractive index (n) of a medium is given by:

$$n = \frac{c}{v} \tag{3.4}$$

where, $c \approx 3 \times 10^8$ m/s, the speed of light in the free space, v is the velocity of light in the medium.

As illustrated in Figure 3.4, when a light ray is incident at the interface of two media (i.e. air and glass), the ray suffers partial refraction and partial reflection. In the following Figure 3.4(a–c), the vertical dotted line represents the normal to the surface. The angles θ_1, θ_2, and θ_r represent the angles that the incident ray, reflected ray, and refracted ray make with the normal.

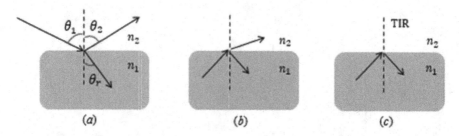

(a) (b) (c)

FIGURE 3.4 (a) A light ray incident on a denser medium ($n_1 > n_2$); (b) a light ray incident from high refractive index medium (i.e. n_1) to the lower refractive index (i.e. n_2); (c) if, the angle of incident is greater than the critical angle, it will undergo total internal reflection.

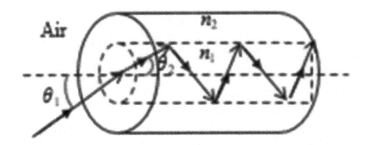

FIGURE 3.5 The guidance of the light in the fiber core through the principle of the total internal reflection.

According to the Snell's law and the law of reflection,

$$n_1 \sin \theta_1 = n_2 \sin \theta_2; \text{ and } \theta_1 = \theta_2 \qquad (3.5)$$

As illustrated in Figure 3.4, when the ray is incident from rarer medium to denser medium, the refracted ray bends towards the normal. On the other hand, if the ray is incident from denser to rarer medium, the refracted ray will bend away from the normal. Further, all the rays including incident, refracted, and reflected lie in the same plane. The angle of incidence for which the angle of refraction is right angle (90°), is known as the critical angle and is denoted by θ_c. In this case, there are no refracted rays in the lower refractive index medium. This phenomenon is known as 'total internal reflection'. Thus, when $\theta_2 = 90°$ the Snell's law becomes

$$\theta_c = \sin^{-1} \frac{n_2}{n_1} \qquad (3.6)$$

When the angle of incident exceeds the critical angle (*i.e.* when $\theta_1 > \theta_c$), there is no refracted ray and total internal reflection takes place. In case of the fiber, when light is incident on one end of the fiber, the light is guided inside the core of the fiber (higher refractive index) through the multiple total internal reflections. The phenomenon of the total internal reflection in the fiber is shown in Figure 3.5. The light is guided in the fiber based on the multiple total internal reflections in the fiber core.

3.4 ACCEPTANCE ANGLE

As shown in Figure 3.6, the light is incident on the input end of the fiber from the air region. The acceptance angle $\left(\theta_a \right)$ is defined as the largest angle for which the refracted ray strikes at the core–cladding interface at critical angle $\left(\theta_a \right)$. The cone made by the incident ray is called the acceptance cone. The cone angle is $2\theta_a$. The rays launched within this cone will guide through the core of the fiber by multiple total internal reflections at the core–cladding interface. However, the light incident outside of this cone is not guiding the core of the optical fiber.

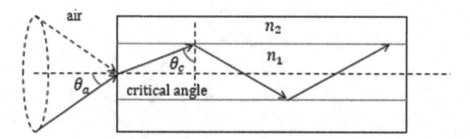

FIGURE 3.6   The ray diagram showing the light launched at the input end of the optical fiber.

3.5 NUMERICAL APERTURE

The light-gathering capacity inside the core of the fiber is known as numerical aperture and is denoted by NA. The rays launched outside the angle specified by a fiber NA will excite the radiation modes of the fiber. Higher is the refractive index of the core with respect to the cladding, meaning larger NA. Numerically, the sine of the acceptance angle of the fiber is called NA.

$$i.e. \text{ NA} = \sin\theta_a$$

$$\text{NA} = \sin\theta_a = \sqrt{\left(n_1^2 - n_2^2\right)} = \theta_1\sqrt{2\Delta} \tag{3.7}$$

where n_1 and n_2 are the refractive index of the core and cladding of the fiber, respectively. The NA of fiber can be determined by measuring the divergence angle of the light cone it emits when all its modes are excited. The NA of the fiber indicates how easily light can be coupled into the fiber.

3.6 V-NUMBER

The V-number is the normalized frequency parameter of the fiber. It is also called the V-parameter of the fiber. The other parameters of the fiber such as the number of modes propagating in the fiber at the specification wavelength, mode-cut-off conditions, and the propagation constant can be written in terms of V. In the mathematical form, the V-number or V-parameter is written as

$$V = \frac{2\pi a}{\lambda} \times \text{NA} = \frac{2\pi a}{\lambda}\sqrt{\left(n_1^2 - n_2^2\right)} \tag{3.8}$$

where, a represents the radius of the core of the fiber, λ is the operating wavelength, n_1 and n_2 are the refractive index of the core and the cladding.

The number of guided modes in the step-index multimode optical fiber is given by

$$\frac{V^2}{2} \tag{3.9}$$

3.7 SINGLE AND MULTIMODE OPTICAL FIBERS

The step-index optical fiber offers single-mode operation when $V < 2.405$. If the value of $V > 2.405$, it offers multimode operation. For silica glass fiber with 1550 nm pumping, the typical diameter of the core of the single-mode fiber is ~9 μm. The multimode silica fiber contains a diameter of more than 9 μm.

You may be curious to ask the question of why optical fibers are made up of glass. According to the Prof. W. A. Gambling, a pioneer in the field of fiber optics, the glass is an outstanding material which has been in its pure form since last more than nine thousand years. There are three most important characteristics of the glass which make it of unparalleled value (i) the viscosity of glass is variable in the wide range of reachable temperatures. In the transition temperature glass can be easily drawn into the thin fiber; (ii) The highly pure silica glass offers very low transmission loss, (iii) The remarkable intrinsic strength of the glass is one of its important characteristics.

3.8 THE PHOTONIC CRYSTAL FIBERS

As discussed in the previous chapter, the PCFs have index guiding and photonic bandgap guiding. The principle of index guiding PCFs is based on modified total internal reflection. While in the case of PCFs with a hollow core (or lower-index core) the guiding principle is based on band-gap guidance. The idea is, that because of the structural property of PCFs, the light waves at certain frequencies reflect back without entering into the core. If there is any defect in the structure of PCF the light starts to propagate through the core of the PCF.

The phenomenon of modified total internal reflection relies on a high index core region (i.e. silica, tellurite, chalcogenides) surrounded by a lower effective index provided by the two-dimensional photonic crystal cladding. The effective refractive index of the two-dimensional photonic crystal cladding of the PCFs can be approximated by standard step-index fiber with a high index core and a lower index cladding. The effective refractive index of the photonic crystal cladding in PCFs depends on the operating wavelength propagating through these PCFs. The effective refractive index of PCF cladding is different from that of the pure silica and depends on the size, shape, arrangement and orientation of the air holes in the cladding region of the PCFs. This behaviour of the effective refractive index of photonic cladding allows PCFs to be designed with a completely new set of properties which are not possible in standard step-index optical fibers. Because of this wavelength-dependent effective refractive index of photonic cladding, it is possible to achieve endlessly single-mode operation, where only a single-mode is supported regardless of the operating wavelength. PCFs can offer single-mode operation for a broad range of the operating wavelengths extending from ultraviolet to far-infrared regions. Therefore, such PCFs are called endlessly single-moded PCFs. Furthermore, it is also possible to alter the dispersion properties of the PCFs based on the shape, size, orientation, and arrangement of the air holes in the cladding region. It is possible to design PCFs with an anomalous dispersion at visible regions. It is also possible to get a flat all-normal nearly zero dispersion profile using PCFs geometries for nonlinear applications such as supercontinuum generation and frequency comb generation.

3.9 SINGLE-MODE CONDITION IN PCFs

In order to estimate whether the PCF is single-mode, the modified normalized frequency, V_{PCF} of the PCF can be simply written as [1]

$$V_{PCF}(v) = k\Lambda \sqrt{\left(n_c^2(v) - n_{cl}^2(v)\right)} \qquad (3.10)$$

where, v is the operating frequency; $n_c(v)$ is the effective index of fundamental mode confined in the core at frequency v, and similarly $n_{cl}(v)$ is the effective index of fundamental space-filling mode which distributes over the cladding with a periodic array of air holes. The single-mode condition for PCF structure is found to be $V_{PCF}(v) < \pi$ [1]. For the V-parameter calculation the numerical values of $n_c(v)$ and $n_{cl}(v)$ can be estimated by the fully-vectorial plane wave method [2]. The numerical values of modified normalized frequency 'V_{PCF}' for telluride PCF with various d/Λ (d represents the diameter of air holes and Λ indicates the distance between the centres of two nearby air holes) at a fixed wavelength of 1.06 μm is illustrated in Figure 3.7 [3]. It is noted from the figure that when $d/\Lambda = 0.5$ the value of V_{PCF} parameter becomes 3.17, which exceeds the limit of single-mode operation (*i.e.* $V_{PCF} = 3.14$). Therefore, the value of d/Λ must be less than 0.5 to ensure single-mode operation.

The PCFs have a much more complex structure in transverse direction compared to that of the conventional step-index optical fibers and have fascinated a great deal of recent interest both in theoretical and experimental aspects. A lot of interest is

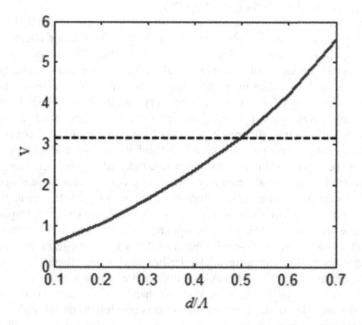

FIGURE 3.7 Variations in numerical value of 'V' parameter with d/Λ of the tellurite PCF
Adapted from Ref. [3].

centred around the additional degree of freedom accessible by the shape, size, orientation, and pattern of the air holes in cladding region. PCFs have revealed interesting landscapes such as broadband single-mode operation [3–5], tailorable effective-modal-areas [6–9], straightforwardly controllable dispersion characteristics [10, 11], and the birefringence [12, 13]. PCFs have shown great promise in nonlinear fiber optics such as four-wave-mixing [14], supercontinuum generation [15–18], and many other novel fiber devices [19–21].

3.10 VARIOUS NUMERICAL METHODS AND THEIR ADVANTAGE AND DISADVANTAGE

When the issue of the numerical modelling has been addressed, it would be natural to consider the methods known for standard waveguide analysis. One of the key points here is to realize the complex nature of the core and the cladding structures of the photonic crystal fiber does generally not allow for the direct use of methods from conventional theory. The denominator of every method is to compute the effective cladding index. Keeping this overall picture in mind and in order to initiate the description of models for PCFs in a manner as close to the generally known approach applied on optical fibers as possible, let us first consider the approximate methods. Since its first demonstration of a PCF (also called a microstructured optical fiber or holey fiber) in 1996 [4], many modelling techniques of PCFs have been reported in their characterization. The names of some modelling techniques are finite element methods, approximate effective index models, finite difference time domain (FDTD) methods, localized-function methods, plane wave expansion methods, multipole methods and finite difference frequency domain (FDFD) techniques.

3.10.1 EFFECTIVE INDEX METHOD

The effective index method was proposed for the first time by Knox *et al.* for the analysis of a rectangular dielectric waveguide structure [22]. In this method, a two-dimensional waveguide structure is transformed to the one-dimensional effective waveguide geometry. Now, this one-dimensional effective waveguide structure can be solved analytically. In case of the photonics crystal fiber, the first step is to calculate the periodically repeated hole-in-silica structure of the cladding and then replace the cladding with a properly chosen effective index. As shown in Figure 3.8, the two-dimensional refractive index profile of the photonics crystal fiber is transformed into its one-dimensional equivalent index profile. In this method, the resulting one-dimensional waveguide then consists of a core and a cladding region that have constant refractive indices n_{co} and n_{cl}, respectively. The core is pure silica, but the definition of the refractive index of the microstructured cladding region is given in terms of the propagation constant of the lowest order mode that could propagate in the infinite cladding material.

The effective index method includes the scalar effect index method (SEIM) and the fully vectorial effective index method (FVEIM). In comparison to SEIM the FVEIM is more accurate. Zhao et al. reported the improved fully vectorial effective index method (IFVEIM) in photonic crystal fiber [23].

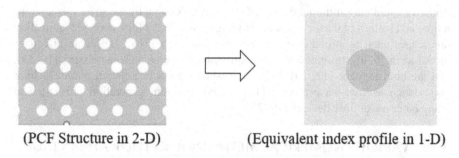

(PCF Structure in 2-D)   (Equivalent index profile in 1-D)

FIGURE 3.8   The schematic representation of the replacement of the transversal cross-section of a PCF structure with an equivalent effective index profile.

The effective index approach to model PCF was first used by Birks et al. [5] and Knight et al. [24]. This method uses weakly guiding approximation [25]. The relative core–cladding index difference,

$$\Delta\left(\approx \frac{n_{co} - n_{cl}}{n_{co}}\right) \leq 1\%,$$

where n_{co} and n_{cl} represent the refractive indices of core and cladding, respectively. More clearly, in the case of index guiding PCF, n_{co} is the refractive index of material (*i.e.* silica) and n_{cl} is the effective refractive index of photonic crystal cladding, known as fundamental space-filling mode (n_{FSM}). As illustrated in Figure 3.9, 2-D structure of PCF cladding is made up of hexagonal unit cells of the air holes. The

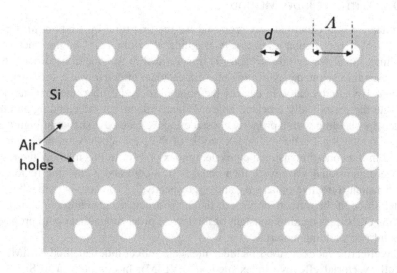

FIGURE 3.9   Cross-sectional view of 2-D triangular photonic crystal lattice with air hole diameter 'd' and pitch 'Λ'.

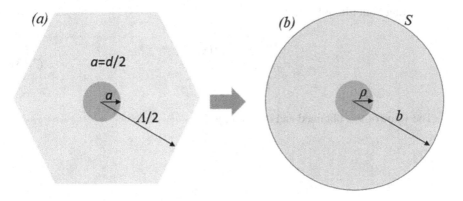

FIGURE 3.10 (a) Hexagonal unit cell of 2D Photonic Crystal having air hole of radius 'a'. (b) Circular approximation of hexagonal unit cell having outer radius '*b*'.

hexagonal unit cell contains an air hole of diameter '*d*' and air-hole spacing of '*Λ*'. In order to obtain the cladding effective index, this hexagonal unit cell can be approximated by a circular unit cell of radius '*b*' as shown in Figure 3.10. The value of '*b*' can be estimated by comparing the air-filling fraction of hexagonal unit cell and circular unit cell. The air-filling fraction (*f*) is given by: *f* = (area of air hole)/(area of unit cell).

The effective index (average index of cladding region) of photonic crystal cladding that has 2-D triangular periodic lattice of air hole in the absence of any core can be estimated. The air holes of diameter '*d*', are separated by a pitch '*Λ*' (i.e. the distance between the centres of two consecutive air holes also called air-hole spacing) is shown in Figure 3.10. These two cladding parameters offer freedom to design the PCF for the requisite purpose, since the effective cladding index shows a strong wavelength dependency in contrast to conventional optical fibers.

The filling fraction of a hexagonal unit cell is defined as the ratio of the area of air hole to the area of hexagonal unit cell, which is given by

$$\frac{\pi \left(\frac{d}{2} \right)^2}{\left(\frac{\sqrt{3}}{2} \right) \Lambda^2} \tag{3.11}$$

The filling fraction for circular unit cell is defined as the ratio of the area of air hole to the area of circular unit cell, which is given by the relation i.e.

$$\frac{\pi \left(\frac{d}{2} \right)^2}{\pi b^2} \tag{3.12}$$

Equating both filling fractions we get, i.e.

$$\frac{\pi\left(d/2\right)^2}{\left(\frac{\sqrt{3}}{2}\right)\Lambda^2} = \frac{\pi\left(d/2\right)^2}{\pi\, b^2} \tag{3.13}$$

The value of b is obtained and is given by,

$$b = \Lambda\sqrt{\left(\frac{\sqrt{3}}{2\pi}\right)} \tag{3.14}$$

The core is pure silica but the effective refractive index of cladding, n_{FSM}, is expressed in terms of the propagation constant, β_{FSM}, of the fundamental space-filling mode (FSM). FSM is defined as the lowest order mode that propagates in the infinite photonic crystal cladding with absence of the core. The effective cladding index, n_{FSM}, is a very important design parameter for realizing a single-mode PCF. β_{FSM} is used to define the effective index of the cladding as,

$$n_{\text{FSM}} = \frac{\beta_{\text{FSM}}}{k_0} \tag{3.15}$$

where, $k_0 = \dfrac{2\pi}{\lambda}$ and λ is the free space wavelength.

3.10.2 Scalar Effective Index Method

In the scalar effective index method, the Scalar wave equation is as follows:

$$\left[\nabla_t + \left(k^2 n^2 - \beta^2\right)\right]\psi = 0 \tag{3.16}$$

where, Δ_t represents the transverse Laplacian operator in cylindrical coordinates, $k = 2\pi/\lambda$, λ indicates the free space wavelength, n is the material index, and β is the propagation constant. By applying the boundary conditions the field must be continuous at the boundary of air hole and silica and it must vanish at the outer boundary of circular unit cell i.e. Ψ and $\left.\dfrac{\partial\Psi}{\partial r}\right|_{R=1}$ must be continuous and $\left.\dfrac{\partial\Psi}{\partial r}\right|_{R=\frac{b}{a}} = 0$, where $R = r/a$, 'a' being the radius of the air hole and 'r' is the normal coordinate to the boundary 'S' and $b = \Lambda\sqrt{\dfrac{\sqrt{3}}{2\pi}}$. Therefore, equations for inner and outer regions of the air hole are given by

$$\psi_1 = A I_0\left(\text{WR}\right) \quad \text{for air hole} \tag{3.17a}$$

$$\psi_2 = B J_0\left(UR\right) + C Y_0\left(UR\right) \text{ Silica region} \tag{3.17b}$$

By applying boundary conditions and making use of Bessel functions, an eigenvalue equation for evaluating the effective index n_{FSM} is given by

$$BJ_1(u) + CY_1(u) = 0 \tag{3.18}$$

where, B and C are the constants and given by the relation

$$B = \frac{A}{J_0(U)}\left[I_0(W) - \frac{WI_1(W)J_0(U) + UJ_1(U)I_0(W)}{U(J_1(U)Y_0(U) - J_0(U)Y_1(U))}\right] \tag{3.19}$$

$$C = \frac{A\left[WI_1(W)J_0(U) + UJ_1(U)I_0(W)\right]}{U\left[J_1(U)Y_0(U) - J_0(U)Y_1(U)\right]} \tag{3.20}$$

with parameters U, W and u as follows

$$\begin{aligned} U &= k_0 a\sqrt{n_s^2 - n_{cl}^2} \\ W &= k_0 a\sqrt{n_{cl}^2 - n_a^2} \\ u &= k_0 b\sqrt{n_s^2 - n_{cl}^2} \end{aligned} \tag{3.21}$$

n_s and n_a are the refractive indices of pure silica and the air, respectively. The modal indices of the FSM are obtained and hence n_{cl} is determined. Subsequently calculating the effective index of cladding, the PCF is assumed to be a step-index fiber as shown in Figure 3.10(b), where the core is pure silica with refractive index n_s of radius 'ρ' surrounded by the uniform cladding of index n_{FSM}. The effective cladding index n_{FSM} for a given PCF can be estimated as described above. The propagation of the lightwave through PCF is analyzed similarly to the step-index fiber. By creating the correlation between the step-index fiber and the PCF, all mathematical formulae of propagation characteristics of the step-index fiber can be applied to the PCF with the core as a silica material and effective cladding index i.e. n_{FSM} for the cladding region. Finally, the field solution for core and cladding in PCF is given as,

$$\Psi = A J_1(U_{eff}R); \quad R < 1 \tag{3.22}$$

$$\Psi = B K_1(W_{eff}R); \quad R > 1 \tag{3.23}$$

where, A and B are the constants that have been determined by applying the boundary conditions. The following eigenvalue equation is obtained for PCF as,

$$\frac{U_{eff}J_1(U_{eff})}{J_0(U_{eff})} = \frac{W_{eff}K_1(W_{eff})}{K_0(W_{eff})} \tag{3.24}$$

where,

$$U_{\text{eff}} = k_0 \rho \sqrt{n_s^2 - n_{\text{eff}}^2} \tag{3.25}$$

$$W_{\text{eff}} = k_0 \rho \sqrt{n_{\text{eff}}^2 - n_{\text{FSM}}^2} \tag{3.26}$$

To obtain the modal parameters of PCF, the core radius of PCF 'ρ' is taken as $\rho = 0.64\Lambda$ [26]. n_s and n_{FSM} represents the refractive indices of pure silica (*i.e.* core) and the cladding (*i.e.* the effective cladding index of PCF). n_{eff} is the effective index of the guided mode.

In the similar way, the normalized propagation constant b_{eff} is given as,

$$b_{\text{eff}} = \frac{n_{\text{eff}}^2 - n_{\text{FSM}}^2}{n_s^2 - n_{\text{FSM}}^2} \tag{3.27}$$

The eigenvalue equation (3.24) has to be solved in order to obtain the effective index of the guided mode.

3.10.3 FULLY VECTORIAL EFFECTIVE INDEX METHOD

In full vectorial effective index method (FVEIM), the effective cladding index method and the effective index of the guided mode of the PCF are estimated by using vectorial equations and hence the accuracy of the results upsurges. The electromagnetic fields in the optical fibers are expressed in cylindrical coordinates as [27]

$$\vec{E} = E(r,\theta)e^{j(\omega t - \beta z)}, \ \vec{H} = H(r,\theta)e^{j(\omega t - \beta z)} \tag{3.28}$$

Substituting into Maxwell's equation we will get the two sets of wave equations.

$$\left[\nabla_t + \left(k^2 n^2 - \beta^2\right)\right]\binom{E_z}{H_z} = 0 \tag{3.29}$$

where, symbols carry their usual meaning. Solving the above Maxwell equation we will get the modal indices of FSM.

$$\left(\frac{P'_1(U)}{UP_1(U)} + \frac{I'_1(W)}{WI_1(W)}\right)\left(n_s^2 \frac{P'_1(U)}{UP_1(U)} + n_a^2 \frac{I'_1(W)}{WI_1(W)}\right) = \left(\frac{1}{U^2} + \frac{1}{W^2}\right)\left(\frac{\beta}{k}\right)^2 \tag{3.30}$$

where, $P_1(U)$ is defined as

$$P_1(U) = J_1(U)Y_1(u) - Y_1(U)J_1(u) \tag{3.31}$$

U, W, u are defined as in Eq. 3.21 and the primes denote differentiation with respect to the argument. In order to calculate the propagation constant β for the PCF, the hexagonal unit cell is approximated by a circular one of radius b and hence the propagation constant of the guided mode will be calculated.

The characteristic equation obtained from the FVEIM is written as

$$\left(\frac{J'_1\left(U_{\text{eff}}\right)}{U_{\text{eff}}J_1\left(U_{\text{eff}}\right)} + \frac{K'_1\left(W_{\text{eff}}\right)}{W_{\text{eff}}K_1\left(W_{\text{eff}}\right)} \right)\left(n_{cl}^2 \frac{J'_1\left(U_{\text{eff}}\right)}{U_{\text{eff}}J_1\left(U_{\text{eff}}\right)} + n_{\text{eff}}^2 \frac{K'_1\left(W_{\text{eff}}\right)}{W_{\text{eff}}K_1\left(W_{\text{eff}}\right)} \right)$$
$$= \left(\frac{1}{U_{\text{eff}}^2} + \frac{1}{W_{\text{eff}}^2} \right)^2 \left(\frac{\beta}{k} \right)^2$$

(3.32)

where parameters U_{eff} and W_{eff} are defined as in the Eqs. 3.25 and 3.26. n_{eff} is the effective index of the fundamental mode and hence n_{cl} is the effective cladding index obtained from Eq. 3.30.

3.10.4 FINITE DIFFERENCE TIME DOMAIN (FDTD) METHOD

The finite difference time domain method was projected by Qiu [28] and Town et al. [29] in 2001 for the full wave analysis of guided modes in PCFs. This method has been successfully used to compute the band structure of photonic crystals, calculation of out-of-plane band structures, defect modes, waveguide modes, and surface modes [30–34]. Since, it is considered that the propagation constant along the propagation direction (z-direction) is fixed for the PCF structures, the two-dimensional mess can be used to calculate the three-dimensional hybrid guided modes.

Formulation

The time-dependent Maxwell's equations for the linear isotropic material in source-free region are given by

$$\frac{\partial H}{\partial t} = -\frac{1}{\mu(r)} \nabla \times E$$

(3.33)

$$\frac{\partial E}{\partial t} = \frac{1}{\varepsilon(r)} \nabla \times H - \frac{\sigma(r)}{\varepsilon(r)} E$$

(3.34)

where, $\mu(r)$, $\varepsilon(r)$, and $\sigma(r)$ represent the position dependent permeability, permittivity, and conductivity of the material, respectively.

We can discretize the Maxwell's equations in space and time using Yee-cell method [35]. In case of the PCF, if the propagation constant β is in the z-direction, each field component will be of the form of $\phi(x,y,z) = \phi(x,y)e^{i\beta z}$, where ϕ represents any field component. Therefore, the z-derivatives can be replaced by $j\beta$ and can be expressed in terms of the transverse variables only.

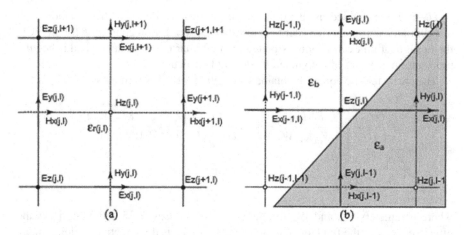

FIGURE 3.11   (a) Yee's 2-D FDTD mesh, (b) 2-D mesh cells across a curved interface.

The unit cell of 2-D mesh over the transverse cross-sectional region of the fibers is represented in Figure 3.11. The discrete form of the x-component of Maxwell's curl equation

$$\frac{\partial H_x}{\partial t} = -\frac{1}{\mu}\left(\frac{\partial E_z}{\partial y} - \frac{\partial E_y}{\partial z}\right)$$ (3.35)

can be written as

$$H_x\Big|_{i,j}^{n+\frac{1}{2}} = H_x\Big|_{i,j}^{n-\frac{1}{2}} - \frac{\Delta t}{\mu_{i,j}}\left(\frac{E_z\big|_{i,j+1}^{n} - E_z\big|_{i,j}^{n}}{\Delta y} - i\beta E_y\big|_{i,j}^{n}\right)$$ (3.36)

where n represents the discrete time-step, and indexes i and j depict the discretized grid point in the x and y-plane, respectively. Δt indicates the increment in time, and Δx and Δy are the intervals between two neighbouring grid points along the x and y-directions respectively. For other field components, the rest of the equations can be achieved in the same manner. To introduce only the real numbers in Eq. (3.36), assume that $E_z, H_x,$ and H_y have components $\cos(\beta z + \phi)$ with real amplitudes, and $H_z, E_x,$ and E_y have components $\sin(\beta z + \phi)$ with real amplitudes, now Eq. (3.36) can be written as for various field components

$$H_x\Big|_{i,j}^{n+\frac{1}{2}} = H_x\Big|_{i,j}^{n-\frac{1}{2}} - \frac{\Delta t}{\mu_{i,j}}\left(\frac{E_z\big|_{i,j+1}^{n} - E_z\big|_{i,j}^{n}}{\Delta y} - \beta E_y\big|_{i,j}^{n}\right)$$ (3.37)

$$H_y\Big|_{i,j}^{n+\frac{1}{2}} = H_y\Big|_{i,j}^{n-\frac{1}{2}} - \frac{\Delta t}{\mu_{i,j}}\left(\frac{E_z\big|_{i,j+1}^{n} - E_z\big|_{i,j}^{n}}{\Delta x} - \beta E_y\big|_{i,j}^{n}\right)$$ (3.38)

$$H_z\Big|_{i,j}^{n+\frac{1}{2}} = H_z\Big|_{i,j}^{n-\frac{1}{2}} + \frac{\Delta t}{\mu_{i,j}} \left(\frac{E_x\Big|_{i,\,j+1}^{n} - E_x\Big|_{i,j}^{n}}{\Delta y} - \frac{E_y\Big|_{i+1,j}^{n} - E_y\Big|_{i,j}^{n}}{\Delta x} \right) \tag{3.39}$$

$$E_x\Big|_{i,j}^{n+1} = \frac{\varepsilon_{i,j} - \sigma_{i,j}\Delta t/2}{\varepsilon_{i,j} + \sigma_{i,j}\Delta t/2} E_x\Big|_{i,j}^{n} + \frac{\Delta t}{\varepsilon_{i,j} + \sigma_{i,j}\Delta t/2} \left(\frac{H_z\Big|_{i,j}^{n+1/2} - H_z\Big|_{i,j-1}^{n+1/2}}{\Delta y} + \beta H_y\Big|_{i,j}^{n+1/2} \right)$$

$$\tag{3.40}$$

$$E_y\Big|_{i,j}^{n+1} = \frac{\varepsilon_{i,j} - \sigma_{i,j}\Delta t/2}{\varepsilon_{i,j} + \sigma_{i,j}\Delta t/2} E_y\Big|_{i,j}^{n} - \frac{\Delta t}{\varepsilon_{i,j} + \sigma_{i,j}\Delta t/2} \left(\frac{H_z\Big|_{i,j}^{n+1/2} - H_z\Big|_{i,j-1}^{n+1/2}}{\Delta x} + \beta H_x\Big|_{i,j}^{n+1/2} \right)$$

$$\tag{3.41}$$

$$E_z\Big|_{i,j}^{n+1} = \frac{\varepsilon_{i,j} - \sigma_{i,j}\Delta t/2}{\varepsilon_{i,j} + \sigma_{i,j}\Delta t/2} E_z\Big|_{i,j}^{n}$$
$$+ \frac{\Delta t}{\varepsilon_{i,j} + \sigma_{i,j}\Delta t/2} \left(\frac{H_y\Big|_{i,j}^{n+1/2} - H_y\Big|_{i-1,j}^{n+1/2}}{\Delta x} - \frac{H_x\Big|_{i,j}^{n+1/2} - H_x\Big|_{i,j-1}^{n+1/2}}{\Delta y} \right) \tag{3.42}$$

For the total number of time steps, the computation time is proportional to the number of discretization points in the computation domain, i.e., the FDTD algorithm is of the order of N. If the following condition is satisfied, the FDTD time-stepping formulas are stable numerically.

$$\Delta t \leq \frac{1}{c\sqrt{\Delta x^{-2} + \Delta y^{-2} + \left(\dfrac{\beta}{2}\right)^2}} \tag{3.43}$$

where c denotes the speed of the light.

At the boundary of FDTD cells, the information out of the computational domain is not available. Therefore, the perfectly matched layer (PML) boundary conditions can be applied for the boundary treatment. In this method, since all the fields are acquired in the time domain, in order to obtain the spectral information we need to perform Fourier transform.

3.10.5   FULL-VECTORIAL FINITE DIFFERENCE FREQUENCY DOMAIN METHOD (FDFD)

The full-vectorial FDFD method can be used to analyze the photonic bandgaps and the mode field profiles of the PCFs with different structures. The FDFD method shares some similarities to the FDTD method. The FDFD method uses Yee's 2-D mesh [35] and an index averaging technique [36]. In the FDFD technique, two discretization schemes are used [36]. One is that first proposed by Stern [37] in which possible discontinuities lie between two adjacent mesh grids and every grid point corresponds to a unique refractive index. This method utterly satisfies the zero divergence conditions to circumvent spurious solutions, and naturally handles physical boundary conditions. The FDFD method provides a neat and compact means of approximating the curl equations with the finite-differences.

3.10.5.1   Formulation

The wave equations for transverse electric field (E_t) and magnetic field (H_t) are given by

$$\left(\nabla_t^2 + k_0^2 \varepsilon_r\right) E_t + \nabla_t \left(\varepsilon_r^{-1} \nabla_t \varepsilon_r . E_t\right) = \beta^2 E_t \tag{3.44}$$

$$\left(\nabla_t^2 + k_0^2 \varepsilon_r\right) H_t + \varepsilon_r^{-1} \nabla_t \varepsilon_r \times \left(\nabla_t \times H_t\right) = \beta^2 H_t \tag{3.45}$$

where $k_0 = 2\pi / \lambda$ represents the wave number in the free space, β is the propagation constant, and ε_r denotes the dielectric constant of the material. The value of β can be directly discretized by finite difference.

The Yee's 2-D mesh is shown in Figure 3.11. For the electric field, the mesh grids lie on the possible dielectric discontinuities. The continuity conditions are automatically satisfied since all the transverse field components are tangential to the unit cell boundaries. Let us consider that the fields are dependent as $\exp\left[i\left(\beta_z - \omega t\right)\right]$.

The Maxwell's curl equations are:

$$\left(\nabla \times E\right) = -\frac{\partial B}{\partial t}, \text{ and } \left(\nabla \times H\right) = -\frac{\partial D}{\partial t}$$

after scaling electric field (E) by the free space impedance $Z_0 = \sqrt{\mu_0 / \varepsilon_0}$, we get

$$ik_0 H_x = -\frac{\partial E_z}{\partial y} - i\beta E_y, \tag{3.46a}$$

$$ik_0 H_y = i\beta E_x - \frac{\partial E_z}{\partial x}, \tag{3.46b}$$

$$ik_0 H_z = \frac{\partial E_y}{\partial x} - \frac{\partial E_x}{\partial y} \tag{3.46c}$$

$$-ik_0 \varepsilon_r E_x = \frac{\partial H_z}{\partial y} - i\beta Y_y \tag{3.47a}$$

$$-ik_0 \varepsilon_r E_y = i\beta H_x - \frac{\partial H_z}{\partial x} \tag{3.47b}$$

$$-ik_0 \varepsilon_r E_z = \frac{\partial H_y}{\partial x} - \frac{\partial H_x}{\partial y} \tag{3.47c}$$

Discretizing Eqs.(3.46) and (3.47) yields

$$ik_0 H_x(j,l) = \frac{\left[E_z(j,l+1) - E_z(j,l)\right]}{\Delta y} - i\beta E_y(j,l), \tag{3.48a}$$

$$ik_0 H_y(j,l) = i\beta E_x(j,l) - \frac{\left[E_z(j+1,l) - E_z(j,l)\right]}{\Delta x}, \tag{3.48b}$$

$$ik_0 H_z(j,l) = \frac{\left[E_y(j+1,l) - E_y(j,l)\right]}{\Delta x} - \frac{\left[E_x(j,l+1) - E_x(j,l)\right]}{\Delta y}, \tag{3.48c}$$

$$-ik_0\varepsilon_{rx}(j,l)E_x(j,l) = \frac{\left[H_z(j,l) - H_z(j,l-1)\right]}{\Delta y} - i\beta H_y(j,l), \tag{3.49a}$$

$$-ik_0\varepsilon_{ry}(j,l)E_y(j,l) = i\beta H_x(j,l) - \frac{\left[H_z(j,l) - H_z(j-1,l)\right]}{\Delta x}, \tag{3.49b}$$

$$-ik_0\varepsilon_{rz}(j,l)E_z(j,l) = \frac{\left[H_y(j,l) - H_y(j-1,l)\right]}{\Delta x} - \frac{\left[H_x(j,l) - H_x(j,l-1)\right]}{\Delta y}, \tag{3.49c}$$

where

$$\varepsilon_{rx}(j,l) = \frac{\left[\varepsilon_r(j,l) + \varepsilon_r(j,l-1)\right]}{2}, \tag{3.50a}$$

$$\varepsilon_{ry}(j,l) = \frac{\left[\varepsilon_r(j,l) + \varepsilon_r(j-1,l)\right]}{2}, \tag{3.50b}$$

$$\varepsilon_{rz}(j,l) = \frac{\left[\varepsilon_r(j,l) + \varepsilon_r(j-1,l-1) + \varepsilon_r(j,l-1) + \varepsilon_r(j-1,l)\right]}{4} \tag{3.50c}$$

The Eqs. (3.50a)–(3.50c) have approximated the refractive indices by averaging the refractive indices of adjacent cells. The matrix form of Eqs.(3.48) and (3.49) is written as

$$ik_0 \begin{bmatrix} H_x \\ H_y \\ H_z \end{bmatrix} = \begin{bmatrix} 0 & -i\beta I & U_y \\ i\beta I & 0 & -U_x \\ -U_y & U_x & 0 \end{bmatrix} \begin{bmatrix} E_x \\ E_y \\ E_z \end{bmatrix}, \tag{3.51}$$

$$-ik_0 \begin{bmatrix} \varepsilon_{rx} & 0 & 0 \\ 0 & \varepsilon_{ry} & 0 \\ 0 & 0 & \varepsilon_{rz} \end{bmatrix} \begin{bmatrix} E_x \\ E_y \\ E_z \end{bmatrix} = \begin{bmatrix} 0 & -i\beta I & V_y \\ i\beta I & 0 & -V_x \\ -V_y & V_x & 0 \end{bmatrix} \begin{bmatrix} H_x \\ H_y \\ H_z \end{bmatrix}, \tag{3.52}$$

where, I represent a square identity matrix, ε_{rx}, ε_{ry}, and ε_{rz} denote the diagonal determined by Eqs. (3.50a)–(3.50c), U_x, U_y, and U_z are the square matrices and depend on the boundary conditions of the rectangular computation window. If a zero-value boundary condition is chosen for the computation window edges, we get

$$
U_x = \frac{1}{\Delta x}
\begin{bmatrix}
-1 & & 1 & & & & \\
& -1 & & 1 & & & \\
& & & & \ddots & & \\
& & & & & \ddots & \\
& & & & & & 1 \\
& & & & & -1 & \\
& & & & & & -1
\end{bmatrix},
$$

$$
U_y = \frac{1}{\Delta y}
\begin{bmatrix}
-1 & & & 1 & & \\
& -1 & & & \ddots & \\
& & \ddots & & & 1 \\
& & & \ddots & & \\
& & & & -1 & \\
& & & & & -1
\end{bmatrix},
$$

$$
V_x = \frac{1}{\Delta x}
\begin{bmatrix}
1 & & & & & \\
-1 & 1 & & & & \\
& -1 & \ddots & & & \\
& & & \ddots & & \\
& & & \ddots & 1 & \\
& & & -1 & & 1 \\
& & & & -1 & 1
\end{bmatrix},
$$

$$
V_y = \frac{1}{\Delta y}
\begin{bmatrix}
1 & & & & & \\
& 1 & & & & \\
-1 & & \ddots & & & \\
& & & \ddots & & \\
& & & & 1 & \\
& -1 & & & & 1
\end{bmatrix},
$$

(3.53)

After some algebra, using Eqs. (3.51) and (3.52) we can get an eigenvalue equation in terms of transverse electric fields:

$$P\begin{bmatrix} E_x \\ E_y \end{bmatrix} = \begin{bmatrix} P_{xx} & P_{xy} \\ P_{yx} & P_{yy} \end{bmatrix} \begin{bmatrix} E_x \\ E_y \end{bmatrix} = \beta^2 \begin{bmatrix} E_x \\ E_y \end{bmatrix}, \tag{3.54}$$

where,

$$P_{xx} = -k_0^{-2} U_x \varepsilon_{rz}^{-1} V_y V_x U_y + \left(k_0^2 I + U_x \varepsilon_{rz}^{-1} V_z \right)\left(\varepsilon_{rx} + k_0^{-2} V_y U_y \right),$$

$$P_{yy} = -k_0^{-2} U_y \varepsilon_{rz}^{-1} V_x V_y U_x + \left(k_0^2 I + U_y \varepsilon_{rz}^{-1} V_y \right)\left(\varepsilon_{ry} + k_0^{-2} V_x U_x \right),$$

$$P_{xy} = U_x \varepsilon_{rz}^{-1} V_y \left(\varepsilon_{ry} + k_0^{-2} V_x U_x \right) - k_0^{-2} \left(k_0^2 I + U_x \varepsilon_{rz}^{-1} V_x \right) V_y U_x, \tag{3.55}$$

$$P_{yx} = U_x \varepsilon_{rz}^{-1} V_x \left(\varepsilon_{rx} + k_0^{-2} V_y U_y \right) - k_0^{-2} \left(k_0^2 I + U_y \varepsilon_{rz}^{-1} V_y \right) V_x U_y,$$

Similarly, in terms of magnetic fields, we can get an eigenvalue equation

$$Q\begin{bmatrix} H_x \\ H_y \end{bmatrix} = \begin{bmatrix} Q_{xx} & Q_{xy} \\ Q_{yx} & Q_{yy} \end{bmatrix} \begin{bmatrix} H_x \\ H_y \end{bmatrix} = \beta^2 \begin{bmatrix} H_x \\ H_y \end{bmatrix}, \tag{3.56}$$

with

$$Q_{xx} = -k_0^{-2} V_x U_y U_x \varepsilon_{rz}^{-1} V_y + \left(\varepsilon_{ry} + k_0^{-2} V_x U_x \right)\left(k_0^2 I + U_y \varepsilon_{rz}^{-1} V_y \right),$$

$$Q_{yy} = -k_0^{-2} V_y U_x U_y \varepsilon_{rz}^{-1} V_x + \left(\varepsilon_{rz} + k_0^{-2} V_y U_y \right)\left(k_0^2 I + U_x \varepsilon_{rz}^{-1} V_x \right),$$

$$Q_{xy} = -\left(\varepsilon_{ry} + k_0^{-2} V_x U_x \right) U_x \varepsilon_{rz}^{-1} V_x + k_0^{-2} V_x U_y \left(k_0^2 I + U_x \varepsilon_{rz}^{-1} V_x \right), \tag{3.57}$$

$$Q_{yx} = -\left(\varepsilon_{rx} + k_0^{-2} V_y U_y \right) U_x \varepsilon_{rz}^{-1} V_y + k_0^{-2} V_y U_x \left(k_0^2 I + U_y \varepsilon_{rz}^{-1} V_y \right),$$

Under the appropriate boundary conditions, we can write:
$V_x = -U_x^T$, and $V_y = -U_y^T$, here the superscript T represents the transpose operation, then from Eqs. (3.56) and (3.57) we have

$$Q_{xx} = P_{yy}^T, Q_{yy} = P_{xx}^T, Q_{xy} = -P_{xy}^T, Q_{yx} = -P_{yx}^T. \tag{3.58}$$

The advantage of this technique is that, once E_t or H_t is obtained, the rest of the field components can be obtained using Eqs. (3.51) and (3.52). The ε_r is real if we do not consider the waveguide material absorption, then P and Q will be real space matrices. Such types of matrices can be stored efficiently in computers to reduce the amount of memory required for the data storage. By solving the eigenvalue equation (3.54) or (3.56) numerically, we can obtain an effective model index, $n_{\text{eff}} \left(= \dfrac{\beta}{k_0} \right)$, and modal fields of the guided modes [36].

3.10.6 FULL VECTORIAL PLANE WAVE EXPANSION METHOD

The plane wave expansion method [38, 39], simple enough to be easily implemented, was used in some of the earliest studies of photonic crystals [40, 41]. This method operates in the reciprocal space. In principle, it is required to calculate the frequencies of allowed modes for all possible k-vectors. However, by utilizing the translational symmetries of the crystals, it is possible to determine any solution (in terms of frequencies) to the eigenvalue problem by only considering k-vectors restricted to 1. This method allows the computation of eigenfrequencies for a photonic crystal structure to any prescribed accuracy, corresponding with the computing time.

3.10.7 MULTIPOLE METHOD

Multipole method is a well-known method for the analysis of electromagnetic scattering problems. This was first introduced by Greengard and Rokhlin [42]. This method can be used to calculate the air-guiding mode in photonic crystal fibers [43]. The multipole method is based on the multipole expansion of the vector Helmholtz equation. The main idea of the multipole method is to treat every single sub-element (or hole) as a scattering element around which the surrounding EM field can be expressed. The significant advantage of the multipole method is its applicability in the prediction of leakage losses in PCFs. In the multipole method, a full vector multipole formulation has been used in which the modes have generally complex propagation constants. The modal fields are expanded in cylindrical harmonic functions about the air holes, which are assumed to be circular.

3.10.8 FOURIER DECOMPOSITION (FD) METHOD

The Fourier decomposition (FD) method is used for the prediction of the confinement loss of silica-air structured optical fibers in which air holes are not necessarily to be circular [44]. In other words, the method is based on the solution of a system of equations, which couples field solutions in the central part of the fiber to outward radiating waves in the outer parts of the fiber.

3.10.9 FINITE ELEMENT METHOD

The finite element method (FEM) is the numerical technique used to execute finite element analysis (FEA) of any given physical phenomenon. FEM was independently developed by engineers to address the structural mechanics problems related to aerospace and civil engineering. The developments began in the mid-1950s with the papers of Turner, Clough, Martin, and Topp [45], and Babuska and Aziz [46]. In this method, the whole domain is subdivided into simpler parts [47, 48]. The simpler part is called the finite element. In this method, the representation of the refractive index profile and mode fields of the profile of the photonic crystal fiber is considered in the first step. In order to obtain a precise description of the field distribution over a fiber cross-section, and especially near the holes, the classical Maxwell differential equations must be solved for a large set of properly chosen elementary subspaces. Doing this it is essential that the conditions of continuity of the fields are taken into account. The general approach first consists of a step, where the cross-section of the modelled guide is split into distinct homogeneous subspaces. This parcelling results in a mesh of simple finite elements: triangles and quadrilaterals in two-dimension case. This approach allows for the application of different-sized elements, depending on the structure and expected mode field response. For a better description of the fields, as the distance of the subspaces to the centre becomes shorter, their dimensions are generally chosen to be smaller. The Maxwell equations are discretized for each element, leading to a set of elementary matrices [49]. The combination of these elementary matrices creates a global matrix system for the entire structure studied. Finally, the effective index and the distributions of the amplitudes and of the polarizations of the modes may be numerically computed. This is done by taking into account the conditions of continuity at the boundary of each subspace.

3.10.10 BEAM PROPAGATION METHOD

The beam propagation method (BPM) is an approximation technique for simulating the propagation of light waves in the slowly varying optical waveguides. The beam propagation method provides a simple intuitive method of determining the modal spectrum and modal profiles for complex waveguides [31]. BPM is a stepwise evaluation of the electromagnetic field through the concatenated sections consisting of equivalent lenses and homogeneous media, also longitudinal variations such as gratings. BPM is a quick and easy method of solving for fields in integrated optical devices. As opposed to scattering problems, BPM is typically used only in solving for intensity and modes within the shaped (i.e. bent, tapered, and terminated) waveguide structures. These structures typically consist of isotropic optical materials, but the BPM has also been extended to be applicable to simulate the propagation of light in general anisotropic materials such as liquid crystals. The BPM method can be used to model bi-directional propagation of the light wave, but the reflections need to be implemented iteratively which can lead to convergence issues.

3.10.11 EQUIVALENT AVERAGED EFFECTIVE INDEX METHOD

The equivalent averaged effective index method was developed by Brechet *et al.* for modelling photonic crystal fibers [26]. In this method, as the name indicates, the index profile is averaged in a manner so that each radial point represents an azimuthal averaging of the refractive index. This method is in good accordance with the finite element method as long as the optical wavelength is not too enlarged.

3.11 DISPERSION CALCULATION

The dispersion is characterized by the broadening of the optical pulse during propagation through the optical fibers. It is the main parameter for designing the optical fiber telecom links. The chromatic dispersion of the single-mode fibers arises due to the dependence of the refractive index of the core and cladding as well as the propagation constant on wavelength. The chromatic dispersion limits the length of the optical fiber links in an optical fiber communication system. In order to overcome this limitation various techniques have been reported to suppress the dispersion effect in optical fiber link. The fiber with negative dispersion can be used in optical fiber link to minimize the dispersion effect. PCFs offer a unique dispersion profile such as large negative dispersion over a wide range of wavelengths. Furthermore, the dispersion in the PCFs can be made normal as well as anomalous by selecting specific design parameters. In this way, the PCF can be a suitable candidate to be used as dispersion compensating devices. The detailed analysis and simulation of propagation characteristics such as dispersion, numerical aperture, spot size, splice loss, and the bend loss of the PCFs by effective index method have been reported [50]. The dispersion property of PCF has been compared by scalar and fully vectorial effective index methods [51].

3.11.1 WAVEGUIDE DISPERSION IN PCFs

Waveguide dispersion in PCFs results from the dependence of modal index on operating wavelength as well as on its geometrical parameters d and Λ.

The waveguide dispersion can be defined as [52]

$$D_w = -\frac{n_2 \Delta}{0.0003} \left(V \frac{d^2(Vb)}{dV^2} \right) \qquad \text{ps / nm / km} \qquad (3.59)$$

For the case of PCF, the above equation can be written as

$$D_{wPCF} = -\frac{n_{cl,eff} \Delta_{eff}}{0.0003} \left(V_{eff} \frac{d^2(V_{eff} b_{eff})}{dV_{eff}^2} \right) \qquad \text{ps / nm / km} \qquad (3.60)$$

The term $\left(V \frac{d^2(Vb)}{dV^2} \right)$ is given be the following relation

$$\left(V \frac{d^2(Vb)}{dV^2} \right) = \frac{2U^2 k_l}{V^2 W^2} \left[\begin{array}{l} (3W^2 - 2k_l(W^2 - U^2)) \\ + W(W^2 + U^2 k_l)(k_l - 1)\left(\dfrac{k_{l-1}(W) + k_{l+1}(W)}{K_l(W)} \right) \end{array} \right] \qquad (3.61)$$

where,

$$k_l = \frac{K_l^2(W)}{K_{l-1}(W) K_{l+1}(W)} \qquad (3.62)$$

The dispersion explicitly depends on the photonic crystal cladding parameters (diameters of the air holes in cladding region (d), the pitch of the air holes (Λ), and the wavelength, λ [50].

3.11.2 Chromatic Dispersion

The dependence of the refractive index of the core material on wavelength must be considered in order to estimate the dispersion characteristic of an optical fiber.

The chromatic dispersion coefficient D is proportional to the second derivative of the second derivative of the modal effective index with respect to the wavelength 'λ' as follows

$$D = -\frac{\lambda}{0.0003}\left(\frac{d^2 n}{d\lambda^2}\right) \tag{3.63}$$

where 'n' represents the effective index of the mode, i.e. the index of the guided fundamental mode. For the PCF the Eq. (3.63) can be written as

$$D = -\frac{\lambda}{0.0003}\left(\frac{d^2 n_{eff}}{d\lambda^2}\right) \tag{3.64}$$

3.12 SPOT SIZE

The spot size is a chief characteristic of an optical fiber. It plays a significant role in the design of the optical communication links based on optical fibers. The mode field diameter, splice losses and bending losses can be determine using the spot size. To determine the spot size of the PCF, the first and second definitions of Peterman spot size is calculated by the following relation

$$w_{p1} = \sqrt{\left[\frac{2\int_0^\infty \psi^2(r)r^2 dr}{\int_0^\infty \psi^2(r)r dr}\right]} \tag{3.65}$$

$$w_{p2} = \sqrt{\left[\frac{2\int_0^\infty \psi^2(r)r dr}{\int_0^\infty \left(\frac{d\psi}{dr}\right)^2 r dr}\right]} \tag{3.66}$$

For the PCF the spot size is given by [50]

$$w_{pcf,I} = \frac{2\rho}{\sqrt{3}}\sqrt{\left[\frac{J_0(U_{eff})}{U_{eff}J_1(U_{eff})} + \frac{1}{W_{eff}^2} - \frac{1}{U_{eff}^2} + 0.5\right]} \tag{3.67}$$

$$w_{pcf,II} = \rho\sqrt{2}\,\frac{J_1(U_{eff})}{W_{eff}J_0(U_{eff})} \tag{3.68}$$

3.13 LOSSES IN OPTICAL FIBERS

3.13.1 SPLICE LOSS

Splice loss occurs because of the misalignment when two optical fibers are joined. It is an important parameter to be evaluated when two PCFs are spliced. The splice loss is calculated by using the spot size. The slice loss due to transverse offset in terms of spot size is represented in the following relation

$$\alpha(dB) = 4.343 \left(\frac{u}{w_{\text{pcf,II}}} \right)^2 \tag{3.69}$$

where, 'u' indicates the transverse offset between two fibers. In the above relation, the Peterman spot size II definition is given.

3.13.2 BEND LOSS

For the application as transmission medium in telecom grade optical fibers it is required to coil the fibers. This will result into bend loss of optical signals in these fibers. To calculate the bend loss of the photonic crystal fiber, the following equation is used [50, 53]

$$\alpha(dB/m) = 4.343 \left(\frac{\pi}{4 \rho_{eq} R_c} \right)^{1/2} \left(\frac{U_{eff}}{V_{eff} K_1(W_{eff})} \right)^2 \left(\frac{1}{W_{eff}} \right)^{3/2} \exp\left\{ -\frac{4 R_c W_{eff}^3 \Delta_{eff}}{3 \rho_{eq} V_{eff}^2} \right\} \tag{3.70}$$

REFERENCES

[1] N. A. Mortensen, J. R. Folkenberg, M. D. Nielsen, and K. P. Hansen, "Modal cutoff and the V parameter in photonic crystal fibers," *Opt. Lett.*, 28, 1879–1881 (2003).

[2] S. G. Johnson and J. D. Joannopoulos, "Block-iterative frequency-domain methods for Maxwell's equations in a planewave basis," *Opt. Express* 8, 173 (2001).

[3] R. K. Sinha, A. Kumar, and T. S. Saini, "Analysis and design of single-mode As$_2$Se$_3$-chalcogenide photonic crystal fiber for generation of slow light with tunable features," *IEEE J. Sel. Topics Quant. Electron.* 22(2), 4900706 (2016).

[4] J. C. Knight, T. A. Birks, P. S. J. Russell, and D. M. Atkin, "All-silica single mode optical fiber with photonic crystal cladding," *Opt. Lett.* 21, 1547–1549 (1996).

[5] T. A. Birks, J. C. Knight, and P. S. J. Russell, "Endlessly single-mode photonic crystal fiber," *Opt. Lett.* 22, 961–963 (1997).

[6] N. G. R. Broderick, T. M. Monro, P. J. Bennett, and D. J. Richardson, "Nonlinearity in holey optical fibers: measurement and future opportunities," *Opt. Lett.* 24, 1395–1397 (1999).

[7] N. A. Mortensen, "Effective area of photonic crystal fibers," *Opt. Express* 10, 341–348 (2002).

[8] T. S. Saini, A. Kumar, and R. K. Sinha, "Triangular-core large-mode-area photonic crystal fiber with low bending loss for high power application," *Appl. Opt.* 53(31), 7246–7251 (2014).

[9] Reena, T. S. Saini, A. Kumar, Y. Kalra, and R. K. Sinha, "Rectangular-core large-mode-area photonic crystal fiber for high power applications: design and analysis," *Appl. Opt.* 55(15), 4095–4100 (2016).

[10] A. Ferrando, E. Silvestre, J. J. Miret, and P. Andres, "Nearly zero ultraflattened dispersion in photonic crystal fibers," *Opt. Lett.* 25, 790–792 (2000).

[11] B. Dabas and R. K. Sinha, "Dispersion characteristic of hexagonal and square lattice chalcogenide As_2Se_3 glass photonic crystal fiber," *Opt. Commun.* 283(7) 1331–1337 (2010).

[12] T. P. Hansen, J. Broeng, S. E. B. Libori, E. Knudsen, A. Bjarklev, J. R. Jensen, and H. Simonsen, "Highly birefringent index-guiding photonic crystal fibers," *IEEE Photon. Technol. Lett.* 13, 588–590 (2001).

[13] T. S. Saini, A. Kumar, and R. K. Sinha, "Large-mode-area single-polarization single-mode photonic crystal fiber: Design and analysis," *Appl. Opt.* 55(19), 4995–5000 (2016).

[14] T. S. Saini, T. H. Tuan, M. Matsumoto, G. Sakai, T. Suzuki, and Y. Ohishi, "Mid-infrared wavelength conversion using dispersion engineered As_2S_5 microstructured optical fiber pumped with ultrafast laser at 2 μm," *Opt. Lett.* 45(10) (2020).

[15] J. K. Ranka, R. S. Windeler, and A. J. Stentz, "Visible continuum generation in air silica microstructure optical fibers with anomalous dispersion at 800nm," *Opt. Lett.* 25, 25–27 (2000).

[16] T. S. Saini, A. Kumar, and R. K. Sinha, "Broadband mid-infrared supercontinuum spectra spanning 2 – 15 μm using As_2Se_3 chalcogenide glass triangular-core graded-index photonic crystal fiber," *IEEE/OSA J. Lightwave Technol.* 33(18), 3914–3920 (2015).

[17] T. S. Saini, A. Bailli, A. Kumar, R. Cherif, M. Zghal, and R. K. Sinha, "Design and analysis of equiangular spiral photonic crystal fiber for mid-infrared supercontinuum generation," *J. Mod. Opt.* 62(19), 1570–1576 (2015).

[18] H. P. T. Nguyen, T. H. Tong, T. S. Saini, L. Xing, T. Suzuki, and Y. Ohishi, "Highly coherent supercontinuum generation in a tellurite all-solid hybrid microstructured fiber pumped at 2 micron," *Appl. Phys. Express* 12(4), 042010 (2019).

[19] B. J. Eggleton, C. Kerbage, P. Westbrook, R. S. Windeler, and A. Hale, "Microstructured optical fiber devices," *Opt. Express* 9, 698–713 (2001).

[20] T. A. Birks, G. Kakarantzas, P. S. J. Russell, and D. F. Murphy, "Photonic crystal fiber devices," *Proc. SPIE 4943, Fiber-based Component Fabrication, Testing, and Connectorization*, 15 April 2003. https://doi.org/10.1117/12.471979

[21] W. Jin, J. Ju, H. L. Ho, et al., "Photonic crystal fibers, devices, and applications," *Front. Optoelectron.* 6, 3–24 (2013). https://doi.org/10.1007/s12200-012-0301-y.

[22] R. M. Knox and P. P. Toulios, "Integrated circuits for the millimeter through optical frequency range," in *Proceeding symposium on submillimeter waves*, J. Fox, ed., Polytechnic Press (Brooklyn, NY, 1970).

[23] X. Zhao, L. Hou, Z. Liu, W. Wang, G. Zhou, and Z. Hou, "Improved fully vectorial effective index method in photonic crystal fiber," *Appl. Opt.* 46(19), 4052–4056 (2007).

[24] J. C. Knight, T. A. Birks, P. S. J. Russell, and J. P. de Sandro, "Properties of photonic crystal fiber and the effective index model," *J. Opt. Soc. Am. A.* 15, 748–752 (1998).

[25] A. W. Snyder and J. D. Love, *Optical Waveguide Theory*, Chapman and Hall (London, 1983), 595–606.

[26] F. Brechet, J. Marcou, D. Pagnoux, and P. Roy, "Complete analysis of the characteristics of propagation into photonic crystal fibers, by the finite element method," *Opt. Fib. Technol.* 6, 181–191 (2000).

[27] K. Okamoto, *Fundamentals of Optical Waveguides*, Academic Press (San Diego, CA, 2000).

[28] M. Qiu, "Analysis of guided modes in photonic crystal fibers using the finite-difference time-domain method," *Microw. Opt. Technol. Lett.* 30(5), 327–330 (2001).

[29] G. E. Town and J. T. Lizier, "Tapered holey fibers for spot size and numerical aperture conversion," *Proc. CLEO'2001*, Paper CtuAA3, P. 261.

[30] C.T. Chan, Q. L. Yu, and K. M. Ho, "Order-N spectral method for electromagnetic waves," *Phys. Rev. B* 51, 16635 (1995).

[31] M. Qiu and S. He, "A nonorthogonal finite-difference time-domain method for computing the band structure of a two-dimensional photonic crystal with dielectric and metallic inclusions," *J. Appl. Phys.* 87, 8268 (2000).

[32] M. Qiu and S. He, "FDTD algorithm for computing the off-plane band structure in a two-dimensional photonic crystal with dielectric or metallic inclusions," *Phys. Lett. A* 278, 348–354 (2001).

[33] M. Qiu and S. He, "Numerical method for computing defect modes in two-dimensional photonic crystals with dielectric or metallic inclusions," *Phys. Rev. B* 61, 12871–12876 (2000).

[34] M. Qiu and S. He, "Guided modes in a two-dimensional metallic photonic crystal waveguide," *Phys. Lett. A* 266, 425–429 (2000).

[35] K. S. Yee, "Numerical solution of initial boundary value problems involving Maxwell's equations in isotropic media," *IEEE Trans. Antenn. Propagat.* 14, 302–307 (1966).

[36] Z. Zhu and T. G. Brown, "Full-vectorial finite-difference analysis of microstructured optical fibers," *Opt. Express* 10(17), 854–864 (2002).

[37] M. S. Stern, "Semivectorial polarized finite difference method for optical waveguides with arbitrary index profiles," *IEE Proc. J. Optoelectron.* 135, 56–63 (1988).

[38] K. Leung and Y. Liu, "Full vector wave calculation of photonic band structures in face-centered-cubic dielectric media," *Phys. Rev. Lett.* 65, 2646–2649 (1990).

[39] Z. Zhang and S. Satpathy, "Electromagnetic wave propagation in periodic structures: Bloch wave solution of Maxwell's equations," *Phys. Rev. Lett.* 65, 2650–2653 (1990).

[40] M. Plihal and A. A. Maradudin, "Photonic band structure of two-dimensional systems: The triangular lattice," *Phys. Rev. B* 44(16), 8565–8571 (1991).

[41] P. R. Villeneuve and M. Piche, "Photonic band gaps in two-dimensional square and hexagonal lattices," *Phys. Rev. B* 46, 4969–4972 (1992).

[42] L. Greengard and V. Rokhlin, A fast algorithm for particle simulations. *J. Comput. Phys.* 73, 325 (1987).

[43] T. White, R. McPhedran, L. Botten, G. Smith, and C. Martijn de Sterke, "Calculations of air-guided modes in photonic crystal fibers using the multipole method," *Opt. Express* 9(13), 721–732 (2001).

[44] L. Poladian, N. Issa, and T. Monro, "Fourier decomposition algorithm for leaky modes of fibers with arbitrary geometry," *Opt. Express* 10(10), 449–454 (2002).

[45] M. J. Turner, R. M. Clough, H. C. Martin, and L. J. Topp, "Stiffness and deflection analysis of complex structures," *J. Aeronaut. Sci.* 23, 805–823 (1956).

[46] I. Babuska and A. K. Aziz, "Survey lectures on the mathematical foundations of the finite element method," in A.K. Aziz, ed., *The Mathematical Foundation of the Finite Element Method with Applications to Partial Differential Equations*, Academic Press (Cambridge, MA, 1972), pp. 3–636.

[47] T. Itoh, G. Pelosi, and P. Silvester, *Finite Element Software for Microwave Engineering*, Wiley-Interscience (New York, 1996).

[48] B. Eggleton, C. Kerbage, P. Westbrook, R. Windeler, and A. Hale, "Microstructured optical fiber devices," *Opt. Express* 9(13), 698–713 (2001).

[49] A. Peyrilloux, S. Fevrier, J. Marcou, L. Berthelot, D. Pagnoux, and P. Sansonetti, "Comparison between the finite element method, the localized function method and a novel equivalent averaged index method for modelling photonic crystal fibers," *J. Opt. A Pure Appl. Opt.* 4, 257–262 (2002).

[50] S. K. Varshney, M. P. Singh, and R. K. Sinha, "Propagation characteristics of photonic crystal fibers," *J. Opt. Commun.* 24, 856 (2003).

[51] R. K. Sinha and A. D. Varshney, "Dispersion properties of photonic crystal fiber: Comparison by scalar and fully vectorial effective index methods," *Opt. Quant. Electron.* 37, 711–722 (2005).

[52] A. Ghatak and K. Thyagarajan, *Introduction to Fiber Optics*, Cambridge University Press (Cambridge, 1999).

[53] A. W. Snyder and J. Love, *Optical Waveguide Theory*, Chapman and Hall (London, 1983), 480–481.

4 Theory of Supercontinuum Generation in Photonic Crystal Fibers

4.1 INTRODUCTION

The dramatic spectral broadening of a very intense short pulse when it passes through nonlinear materials such as fibers and planar waveguides is known as supercontinuum generation (SCG). This phenomenon was first observed by *Alfano and Shapiro* in 1970 using solid and gaseous nonlinear media [1]. The schematic representation of the principle of SCG is illustrated in Figure 4.1. The SCG is the result of the combined response of various nonlinear effects (*i.e.* four-wave mixing, self-phase modulation, cross-phase modulation, and stimulated Raman scattering) along with the dispersion characteristic of the fiber. If we talk about the optical fibers, the supercontinuum was first demonstrated by *Lin and Stolen* using nanosecond laser pulses of dye laser [2]. Since the beginning of the twenty-first century, the SCG has been reported in various specialty optical fibers i.e. photonic crystal fibers (PCFs) or microstructured optical fibers (MOFs) [3–9]. The supercontinuum spectrum extending from 2 µm to 15 µm was first numerically reported in 2015 using photonic crystal fiber in chalcogenide material [9]. After that, the supercontinuum spectrum ranging from 2 µm to 15.1 µm was first demonstrated experimentally in 2016 using chalcogenide step-index optical fiber [10].

During the last two decades, SCG in optical fibers has appeared as a dynamic and exhilarating research field. Various kinds of new light sources have been formed which are finding potential applications in diverse fields including optical communications [11], optical coherence tomography [12–14], fluorescence lifetime imaging [15], absorption spectroscopy [16], atmospheric gas sensing [17], food quality analysis [18], early cancer diagnostics [19], and biomedical endoscopy [20, 21]. Supercontinuum light is generated when a large number of nonlinear phenomena

Laser pulse Nonlinear PCF SCG

FIGURE 4.1 The schematic of the spectral broadening of intense light pulse after passing from nonlinear photonics crystal fiber.

 DOI: 10.1201/9781003502401-4

such as self-phase modulation (SPM), stimulated Raman scattering (SRS), cross-phase modulation (XPM) and four-wave mixing (FWM) act together upon a pump beam in order to cause wide-ranging spectral broadening of the original pump pulse. The developments in recent years have elevated scientific anxieties to understand and model how all these nonlinear processes interact together to generate supercontinuum light and how the geometrical parameters of fiber can be engineered to regulate and improve the supercontinuum spectral broadening.

The mechanism of the spectral broadening of the supercontinuum spectrum primarily depends on the dispersion characteristic of the optical fiber structure and the input pulse property. When a highly intense laser pulse is incident on the nonlinear optical medium, in the anomalous group velocity dispersion regime, it evolves towards the higher-order solitons [22]. The higher-order solitons are affected by the higher-order dispersion and SRS in the case of femtosecond pumping. Subsequently, the higher-order solitons become unsteady and start to break up into several fundamental solitons through the fission process. Such messy soliton fission progression causes shot-to-shot noise in the supercontinuum spectrum [23, 24]. But in the case of pumping in the normal group velocity dispersion regime, for femtosecond pulses, SPM is only the reason for ultra-broadband spectral broadening [24]. Such spectral broadening in the supercontinuum spectrum is suitable for time-resolved applications including optical coherence tomography, nonlinear microscopy, and pump-probe spectroscopy. One of the possible techniques for generating supercontinuum spectrum, in the normal dispersion regime, is to pump the fiber far below the zero dispersion wavelength (ZDW), so that the generated spectrum does not extend into the anomalous dispersion region. However, this would require a very high power input laser or very short laser pulses to overcome the short effective interaction length due to the high value of the dispersion. Though, the possibility of broadening supercontinuum spectra beyond 20 μm was expected using the tellurium-based chalcogenide glasses [25].

4.2 NONLINEAR MECHANISM INVOLVED IN THE SUPERCONTINUUM BROADENING

In the case of picosecond laser pumping, SPM alone can yield considerable spectral broadening in the pulse spectrum after passing through nonlinear optical fibers. The spectral broadening factor is given by the equation [22]

$$\delta\omega_{max} = 0.86\Delta\omega_0\phi_{max} \tag{4.1}$$

with $\Delta\omega_0$ being the 1/e half-width. One can calculate the maximum SPM-induced phase shift by the relation $\phi_{max} = \gamma P_0 L_{eff}$; where γ is the nonlinear coefficient, P_0 represents the input power of the pulse, and L_{eff} indicates the effective length of the fiber used. For example, in the case of input picosecond pulse with the spectral width of ~1 nm, the calculated spectral broadening factor is ~10, the SPM alone cannot produce the supercontinuum spectrum extending over 100 nm or larger. Another nonlinear mechanism to produce new wavelengths is the SRS. For the high input peak power P_0, the SRS creates a Stokes band on the longer side of the spectrum shifted by ~13 THz from the centre of the pulse spectrum. In this way, SRS affects the supercontinuum spectrum by improving it selectively on the long-wavelength side. Hence, makes the supercontinuum spectrum asymmetric. Another nonlinear effect is the FWM which can generate new frequency components on both sides of the pulse spectrum. To the FWM phenomena, the phase matching condition must be satisfied.

It is worthwhile to mention here that, for the larger spectral bandwidth in super-continuum, the group velocity dispersion (GVD) parameter β_2 is to be taken as wavelength-dependent in any theoretical modelling. The fiber geometries in which the dispersion property is allowed to change along the fiber length, the supercontin-uum process can be improved. The numerical simulations suggest that the flatness of the supercontinuum spectrum improved considerably if the parameter β_2 increases along the length of the fiber [26]. For the pump wavelength of 2000 nm, a 2030 nm-wide coherent supercontinuum spectrum could be generated using a 3.2 cm long tapered tellurite fiber in which dispersion is not only varied along the length of the fiber but also remain all-normal throughout the whole range of the generated super-continuum [27]. One of the paybacks of the tapered design of the fiber is that the length of the fiber for generating SCG can be reduced significantly. The all-normal, nearly zero, and flat dispersion characteristic of the fiber is very important to obtain coherent SCG. In the 2020 experiment, the coherent mid-IR supercontinuum light was demonstrated using an all-normal dispersion engineered short tapered chalco-genide fiber pumped with fs laser pulses at various wavelengths ranging from 2 μm to 2.6 μm [28]. The nearly symmetric nature of the generated supercontinuum spec-tra specifies that the Raman gain played a relatively negligible role. The combination of SPM, XPM, and FWM is responsible for most of the spectral broadening.

4.3 NUMERICAL MODELLING OF THE SUPERCONTINUUM SPECTRUM

When an electromagnetic field propagates in the medium, it interacts with the atoms of the medium, and consequently pulse experiences loss and dispersion. The disper-sion effect arises because the different wavelength components of the pulse move at different velocities due to the wavelength dependence of the refractive index. The total dispersion in the optical waveguide has an additional component due to the light confinement known as waveguide dispersion. In the case of high field intensity, the medium also reacts in a nonlinear model and the refractive index becomes intensity-dependent and photons can interact with the phonons (molecular vibrations) of the medium. The nonlinear propagation equations can be derived in time as well as in the frequency domain. The time-domain formulation is preferred because of its analytic similarity with the nonlinear Schrodinger equation. However, the frequency-domain formulation does show more directly the frequency-dependence of the effects such as loss, dispersion, and effective-mode-area.

The linearly x-polarized electric field is given by the relation $E(r,t) = \frac{1}{2}x\left[E(x,y,z,t)\exp(-iw_0 t) + c.c.\right]$. In the frequency domain, the Fourier transform of the electric field $E(x,y,z,t)$ is given by $\tilde{E}(x,y,z,\omega) = F(x,y,\omega)\tilde{A}(z,\omega - \omega_0)\exp(i\beta_0 z)$, where $\tilde{A}(z,\omega)$ is the complex spectral envelope, ω_0 is the reference frequency (the centre frequency of the input pulse), β_0 represents the wavenumber at the centre frequency of the input pulse. $F(x, y, \omega)$ represent the transverse modal distribution. The time-domain envelope is calculated from the following relation

$$A(z,t) = \mathcal{F}^{-1}\{\bar{A}(z,\omega-\omega_0)\} = \frac{1}{2\pi}\int_{-\infty}^{\infty} \bar{A}(z,\omega-\omega_0)\exp\left[-i(\omega-\omega_0)t\right]d\omega, \quad (4.2)$$

where the amplitude is normalized such that $|A(z, t)|^2$ provides the instantaneous power (in watts) and $\mathcal{F}^{-1}$ represents the inverse Fourier transform.

Using the above definition of inverse Fourier transform and implementing the change of variable $T = t - \beta_1 z$ to transform into a co-moving frame at the envelope group velocity $\dfrac{1}{\beta_1}$, one can get a time-domain generalized nonlinear Schrodinger equation (NLSE) for the evolution of $A(z, t)$ [29]

$$\frac{\partial A}{\partial z} + \frac{\alpha}{2}A - \sum_{k\geq 2}\frac{i^{k+1}}{k!}\beta_k\frac{\partial^k A}{\partial T^k}$$
$$= i\gamma\left(1+i\tau_{\text{shock}}\frac{\partial}{\partial T}\right)\times\left\{A(z,T)\int_{-\infty}^{\infty}R(T')\left|A(z,T-T')\right|^2 dT'\right\}$$
(4.3)

In the above equation, the left-hand side models linear propagation effects, α represents the linear power attenuation, and the β_k is the dispersion coefficients associated with Taylor series expansion of the propagation constant $\beta(\omega)$ about the reference frequency ω_0. The dispersion operator can be applied directly and in an approximation-free manner in the frequency domain through multiplication of the complex spectral envelope $\bar{A}(z,\omega)$ by the operator $\bar{\beta}(\omega) = (\beta(\omega)-(\omega-\omega_0)\beta_1 - \beta_0)$. In the right-hand side is the nonlinear coefficient

$$\gamma\left(=\frac{\omega_0 n_2(\omega_0)}{cA_{\text{eff}}(\omega_0)}\right)$$
(4.4)

with $n_2(\omega_0)$ represents the nonlinear refractive index at frequency ω_0, c denotes the velocity of light, and $A_{\text{eff}}(\omega_0)$ represents the effective-mode-area at frequency at ω_0. $R(t)$ represents the Raman response function which is given by the relation:

$$R(t) = (1-f_R)\delta(t) + f_R h_R(t)$$

or

$$R(t) = (1-f_R)\delta(t) + f_R\frac{\tau_1^2+\tau_2^2}{\tau_1\tau_2^2}\exp\left(-\frac{t}{\tau_2}\right)\sin\left(\frac{t}{\tau_1}\right)H(t)$$
(4.5)

where f_R denotes the fractional contribution of the Raman response, $\delta(t)$ represent the Dirac delta function, τ_1 is the Raman period, τ_2 provides the damping time of the network of vibrating atoms, and H(t) represents the Heaviside step function (H(t) = 0 for t < 0 & H(t) = 1 for t > 0). The time derivation term models the dispersion of the nonlinearity. This term is associated with effects such as self-steepening and optical shock formation, characterized by a timescale $\tau_{\text{shock}} = \tau_0 = 1/\omega_0$. For the case of fiber propagation, additional dispersion of the nonlinearity arises from the frequency dependence of the effective area, and τ_{shock} can be generalized to account for this in an approximate manner.

A better approach to include the dispersion of the nonlinear response is to describe it directly in the frequency domain. We can derive a generalized NLSE (GNLSE) for the pulse evolution as:

$$\frac{\partial \tilde{A}'}{\partial z} = i\bar{\gamma}(\omega)\exp\left(-\hat{L}(\omega)z\right)\mathcal{F}\left\{\bar{A}(z,T)\int_{-\infty}^{\infty} R(T')\left|\bar{A}(z,T-T')\right|^2 dT'\right\} \qquad (4.6)$$

where $\tilde{A}'$ represents the envelope of an output pulse in the frequency domain which is related to the envelope of the pulse in the time-domain by the following equation

$$\bar{A}(z,T) = \mathcal{F}^{-1}\left\{\frac{\tilde{A}(z,\omega)}{A_{\text{eff}}^{\frac{1}{4}}(\omega)}\right\} \qquad (4.7)$$

where $\mathcal{F}^{-1}$ represents the inverse Fourier transform, z shows the propagation distance, and A_{eff} indicates the frequency-dependent effective-mode-area of the mode propagating in the fiber.

$\bar{\gamma}(\omega)$ indicates the frequency-dependent nonlinear coefficient and is given by the following equation

$$\bar{\gamma}(\omega) = \frac{n_2 n_0 \omega}{c n_{\text{eff}}(\omega) A_{\text{eff}}^{1/4}(\omega)} \qquad (4.8)$$

where n_2 is the nonlinear refractive index ($n_2 = 2.3 \times 10^{-17}$ m²/W for AsSe$_2$ based chalcogenide glass [30]), n_0 represents the linear refractive index of the glass at the wavelengths used to determine n_2, c indicates the velocity of the light in the vacuum, and n_{eff} is the frequency-dependent effective refractive index of the propagating mode in the fiber. It is to be noted here that the frequency-dependent nonlinear coefficient $\bar{\gamma}(\omega)$ defining in the Eq. (3.8) has a different unit than the standard definition of the nonlinear coefficient γ mentioned in the Eq. (3.4). It is advisable not to compare directly these two definitions of nonlinear coefficients. This inconsistency will be resolved after rewriting the GNLSE in the form of the Eq. (3.13).

The change of the variables is given by the relation

$$\tilde{A}'(z,\omega) = \tilde{A}(z,\omega)\exp\left(-\hat{L}(\omega)z\right) \tag{4.9}$$

where $\hat{L}(\omega)$ indicates the linear operator which is given by the relation:

$$\hat{L}(w) = i\left(\beta(\omega) - \beta(\omega_0) - \beta_1(\omega_0)\left[\omega - \omega_0\right]\right) - \frac{\alpha(\omega)}{2} \tag{4.10}$$

where β represents the propagation constant, β_1 is the reciprocal of the group velocity of the envelope, ω_0 depicts the reference frequency, and α indicates the losses in the fiber. Because of the change of the variable mentioned in Eq.(3.9), the Eq.(3.6) is an ordinary differential equation (ODE) which can directly be integrated using standard ODE codes. To understand how this is related to the time-dependent Eq.(3.3), and also to get the form of GNLSE with the definition of γ which can be directly compared with the standard definition given in Eq.(3.4). After inserting Eqs. (3.7) and (3.9) in the Eq.(3.6), we get

$$\frac{\partial \bar{C}}{\partial z} - i\bar{\beta}(\omega)\tilde{C}(z,\omega)$$

$$= i\frac{n_2 n_0 \omega}{cn_{\text{eff}}(\omega)\sqrt{A_{\text{eff}}(\omega)A_{\text{eff}}(\omega_0)}}\mathcal{F}\left\{C(z,t)\int_{-\infty}^{\infty}R(T')\left|C(z,T-T')\right|^2 dT'\right\}; \tag{4.11}$$

where

$$\mathcal{F}\left\{C(z,t)\right\} = \tilde{C}(z,\omega) = \left[\frac{A_{\text{eff}}(\omega)}{A_{\text{eff}}(\omega_0)}\right]^{-1/4}\tilde{A}(z,\omega) \tag{4.12}$$

Let us define the frequency-dependent nonlinear coefficient as

$$\gamma(\omega) = \frac{n_2 n_0 \omega_0}{cn_{\text{eff}}(\omega)\sqrt{A_{\text{eff}}(\omega)A_{\text{eff}}(\omega_0)}}; \tag{4.13}$$

The Eq.(3.11) becomes

$$\frac{\partial \tilde{C}}{\partial z} - i\left[\beta(\omega) - \beta(\omega_0) - \beta_1(\omega_0)\{\omega - \omega_0\}\right]\tilde{C}(z,\omega) + \frac{\alpha(\omega)}{2}\tilde{C}(z,\omega)$$

$$= i\gamma(\omega)\left[1 + \frac{\omega - \omega_0}{\omega_0}\right]\mathcal{F}\left\{C(z,t)\int_{-\infty}^{\infty}R(T')\left|C(z,T-T')\right|^2 dT'\right\}; \tag{4.14}$$

The above mentioned GNLSE [i.e. Eq.(3.14)] has the advantages as follows:

(i) The definition of $\gamma(\omega)$ reduces to the standard definition Eq.(3.4) if we neglect the frequency dependencies of A_{eff} and n_{eff}.

(ii) If we neglect the frequency dependencies and expanding $\beta(\omega)$ as a Taylor series and inverse-Fourier transforming Eq.(3.14) leads to the well-known time-domain equation written as Eq.(3.3).

The Eq. (3.3), (3.6) and (3.14) have already been used to model the dynamics of nonlinear pulse propagation and fiber SCG over a wide parameter range [3]. It is worthwhile to mention here that the input pulse noise and the forcing noise due to the spontaneous Raman scattering may be added to the equations [31]. The statistical properties of the supercontinuum spectrum can be determined by the numerical simulations based on these equations. The stochastic simulations and measured RF noise properties of the supercontinuum spectra were thoroughly compared by Corwin et al. [32]. The coherence characteristic of the generated supercontinuum spectrum is affected by the existence of the quantum noise of the pulse. The one-photon-per-mode semi-classical theory can be used to model the noise property of the input pulse [30]. The complex degree of coherence can be used to consider the deficit in the coherence characteristic of the spectrum due to the spectral phase instability at each wavelength. The relation for the complex degree of coherence is given by the relation [3]

$$\left| g_{12}^{(1)} \left(\lambda, t_1 - t_2 \right) \right| = \left| \frac{\left\langle E_1^* \left(\lambda, t_1 \right) E_2 \left(\lambda, t_2 \right) \right\rangle}{\sqrt{\left\langle \left| E_1 \left(\lambda, t_1 \right) \right|^2 \right\rangle} \sqrt{\left\langle \left| E_2 \left(\lambda, t_2 \right) \right|^2 \right\rangle}} \right| \tag{4.15}$$

where E_1 and E_2 are the amplitudes of the electric field for two successive generated spectra. To focus on the wavelength dependence of the coherence, in the simulation study $t_1 - t_2 = 0$ can be taken.

It is important to discuss here about the numerical techniques used to solve the GNLSE. The split-step Fourier method is the pillar of the numerical methods in the nonlinear fiber optics. It falls in the category of pseudo-spectral methods which are normally employed to contract with the stiff differential equations having diffractive or dispersive terms. In order to solve linear and nonlinear terms in Eq. (3.3) or Eq. (3.14) in a natural and efficient way, this method involves alternating between the time- and frequency-domains. The linear term can be completely integrable, so this term of the equation can be exactly accounted for in each step. The nonlinear term is generally integrated using Runge–Kutta methods. The Eq. (3.6) can be integrated directly using Runge–Kutta techniques because the stiff dispersive part has been removed due to the change of variables.

4.4 PHYSICAL MECHANISMS INVOLVED IN THE GENERATION OF SUPERCONTINUUM LIGHT

The SCG in optical fiber comprises the interplay between the linear and nonlinear effects that happen during the propagation of the laser pulses or continuous laser waves through the fiber. Though the complex dynamics of this interaction can some-times make it challenging to recognize contributing mechanisms in segregation, an understanding of how different effects act independently is nevertheless essential in developing an understanding of physics of the spectral broadening process. Some linear and nonlinear effects that are responsible for the broadening of the intense laser pulse are discussed below.

4.4.1 DISPERSION

Dispersion comes in the category of linear optical effects. It plays a critical role in influencing the characteristic of nonlinear interactions in a fiber. It arises due to the frequency variation of the effective index of the guided mode. The dispersion effect depends on both the material and the waveguide contributions. The disper-sion property of the conventional step-index optical fibers with cylindrical symmetry can be determined analytically. However, the dispersion character of the specialty optical fibers such as PCFs requires numerical computation. In addition to the chro-matic dispersion, dispersion is due to polarization mode dispersion in the case of a birefringent fiber or the higher order intermodal dispersion between transverse modes in the multimode fibers. It is important to mention here that the dispersion impacts both the phase and group velocities of an optical signal propagating into the optical fibers. In the investigation of the nonlinear frequency conversion processes the phase-matching criteria play a vital role. Therefore, in general, both the linear and nonlinear contributions to overall phase mismatch must be taken into account. However, to limit the interaction length of the fiber the group velocity mismatch also plays a very important role.

The group velocity dispersion (GVD) parameter β_2 and the higher-order disper-sion terms can be defined using the Taylor series expansion of the propagation con-stant. There are two complementary definitions of GVD used in the literature: (i) β_2 (in the unit of s^2m^{-1}), generally used in physics and the ultrafast photonics; (ii) $D = -(2\pi c/\lambda^2)\beta_2$ (in the unit of ps $\times$ nm^{-1}km^{-1}), generally employed in engineer-ing fields. The wavelength at which $D = 0$ (or $\beta_2 = 0$) is known as zero dispersion wavelength (ZDW). The range of the wavelength for which $D < 0$ (or $\beta_2 > 0$) is referred to as the normal dispersion region. However, the range of the wavelength for which $D > 0$ (or $\beta_2 < 0$) is called anomalous dispersion regime. The ZDW depends on the fiber geometry. Some specialty fibers can offer multiple ZDWs. An initially-unchirped input laser pulse propagating in linear media will temporally broaden in an identical way irrespective of the sign of β_2. Nonetheless, in the case of the input laser pulse propagating in a nonlinear medium, the sign of the dispersion (normal or anomalous) impacts the propagation dynamics considerably.

Saitoh et al. reported a new controlling technique based on the full-vector finite element method with anisotropic perfectly matched layers for controlling the chro-matic dispersion in the index-guiding PCFs [33]. The photonic crystal fiber with four rings of air holes can provide flattened dispersion with a variation of 0 ± 0.5 ps/nm/km in the spectral range of 1.19 μm to 1.69 μm. However, the photonic crystal fiber with five rings of air holes can provide flattened dispersion of 0 ± 0.4 ps/nm/km in the spectral range of 1.23 μm to 1.72 μm [33]. Hsu et al. reported numerically a dis-persion compensating photonic crystal fiber design for the spectral ranging from

1338 nm to 1564 nm covering E, S, and C bands [34]. The nano-core photonic crystal fiber with air holes arranged spirally in the cladding region can be designed for negative dispersion and high nonlinearity at 1.55 μm [35].

4.4.2 SELF AND CROSS-PHASE MODULATION

The self-phase modulation (SPM) and cross-phase modulation (XPM) come in the category of nonlinear effects. If we consider the Kerr nonlinearity of the fiber and neglect the dispersion effects, then SPM and XPS will be obtained. These effects (*i.e.* SPM & XPM) arise due to the time-dependent intensity profile of the intense laser pulse which causes the modulation to the local refractive index of the fiber material. The intensity-dependent refractive index is given by the relation: $\Delta n = n_2 I(t)$. This arises, in a time-dependent phase delay to the same pulse in the case of SPM or a co-propagating pulse in the case of XPM. Consequently, a corresponding nonlinear chirp with the generation of new frequency components arises. If one considers SPM in seclusion for a temporally symmetric unchirped input laser pulse, the broadening of the pulse spectrum is symmetric and the time-domain envelope of the pulse is unaffected. The interplay between dispersion and nonlinear effects is also important in the spectral broadening of an input laser pulse. Dispersion behaves differently in normal and anomalous regimes of the electromagnetic spectrum. The normal dispersion and the SPM intermingle to lead to simultaneous temporal and spectral broadening of the spectrum with the growth of a linearized chirp across the central region of the pulse. This growth, under some specific situations, can lead to unusual self-similar propagation mechanisms. Another broad range of interaction between the SPM and dispersion ascends in the anomalous regime. The interaction of SPM and the anomalous dispersion causes the dynamics lead to the generation of the solitons. In the case of the XPM, the dynamics of the generation of supercontinuum can be somewhat multifarious as two interacting pulses can be propagating in normal and anomalous regimes of the dispersion. An efficient interaction happens between the wavelength components where there is a low group velocity mismatch. XPM is shown to have a crucial significance as it extends the supercontinuum spectrum towards the shorter wavelengths.

4.4.3 SOLITONS

In the case of anomalous dispersion regime of the optical fiber, the nonlinear chirp arises by the SPM and the linear chirp induced by GVD can associate with each other to produce either stable or periodic progression of the optical solitons. The solitons are the analytic solutions of the nonlinear Schrodinger equation (NLSE) shown above in Eq. 3.3 without taking higher-order dispersion and the Raman terms. Based on the initially injected pulses there are various kinds of solitons. The order of soliton can be estimated by the pulse and fiber parameters using the relation: $N^2 = L_D/L_{NL}$, where, $L_D = \dfrac{T_0^2}{|\beta_2|}$ (T_0 being the pulse width) is the dispersion length, and $L_{NL} = \dfrac{1}{\gamma P_0}$ (P_0 is the peak power of the input pulse) represents the nonlinear length scale. The fundamental soliton corresponding to N = 1. For fundamental soliton both the temporal and spectral profiles remain unaffected during the propagation. The higher order solitons are corresponding to $N \geq 2$. The higher-order solitons experience periodic temporal and spectral evolution during the propagation. Though the general form of the evolution can be very complex. It is because of the fiber anomalous GVD, the initial phase of propagation is always associated with simultaneous nonlinear spectral broadening and temporal compression.

During the supercontinuum process, originated by an input laser pulse experiencing normal GVD with negative dispersion slop, the Raman soliton exhibits unique dynamics. The blue components of the spectrum form a Raman soliton that moves faster than the input optical pulse and ultimately decelerates due to Raman-induced frequency downshifting. In the time-domain, the soliton route bends and comes to be vertical when the Raman shift stops to occur as the spectrum of Raman soliton approaches the zero dispersion point. Parts of the red components of the pulse spectrum are grabbed by the Raman soliton through cross-phase modulation and they travel with it [36].

4.4.4 SOLITON PERTURBATION AND DISPERSIVE WAVES

When highly intense laser pulses are launched into an optical fiber the anomalous dispersion regime of the fiber leads to soliton dynamics. In fact, the ideal soliton propagation in optical fibers is obtained only under extremely strict conditions. To understand the mechanism of the generation of supercontinuum spectrum in optical fibers, the knowledge of some propagation effects under soliton dynamics is essential. In general, an intense laser pulse incident in the anomalous dispersion region of the fiber is associated with the soliton number, $N \gg 1$. In the case of $N \gg 1$, even when N is non-integral, the initial stage of evolution of spectral broadening will be qualitatively same as that of a higher order soliton in that there will be an initial phase of the temporal compression and the spectral broadening. Nonetheless, the higher-order solitons (i.e. $N \gg 1$) are inherently unstable. Hence, the existence of any perturbation willingly breaks up or decays into a series of lower amplitude subpulses. The process of decaying of the higher order solitons into the lower amplitude pulses is known as soliton fission. In the case of femtosecond pulse regime, the most significant effects that can disturb the growth of the higher-order soliton and induce pulse break-up are the Raman scattering and higher-order dispersion effects. The length of the fiber at which the fission process ensues customarily corresponds to the point where the emerging input pulse soliton extends its maximum bandwidth. This length of the fiber is referred to by the 'fission length', and written by the relation: $L_{fiss} = \dfrac{L_D}{N}$. It is to be cconcluded here that the basis of the higher-order soliton effect compression technique is to optimize the length of the fiber just shorter than the soliton fission length. In addition to the soliton perturbation, there are another two main effects crucial to the generation of additional bandwidth beyond the initial spectral broadening phase of the incident pulse. A soliton which is evolving close to the ZDW can transfer a part of its energy to the normal GVD regime to produce what is variously referred to as a 'non-solitonic radiation', 'dispersive wave, or Cherenckov radiation. The non-solitonic radiation or dispersive wave was first numerically reported by Wai et al. in a single-mode optical fiber [37]. The generation of dispersive wave can be analyzed in terms of the phase-matching conditions which involve the soliton linear and nonlinear phase and the linear phase of a continuous wave at an altered frequency [38]. The solitons, because of the soliton self-frequency shift, can experience continuous displacement to longer wavelengths even if perturbed by the generation of dispersive wave. This is due to the individual bandwidths of the solitons overlap the Raman gain. The frequency shift can be expressed by the relation: $\frac{dv_R}{dz} \propto \frac{|\beta_2|}{T_0^4}$ [39]. However, the dispersion of the nonlinear response can slow-down this frequency shift as the solitons shift to the longer wavelengths [40]. These effects arise concurrently, and the interaction between them can be subtle. Additionally, the effects of the cross-phase modulation and the group velocity matching between the dispersion wave and the soliton can induce trapping dynamics and lead to further enhance in the bandwidth on both the short and long spectral edges of the supercontinuum spectrum [41–43]. The soliton plays a very important role in directional SCG. Numerical simulation indicates that in a silicon nitride waveguide with an anomalous dispersion region surrounded by two normal-dispersion regions and pumped in one of the normal dispersion regions, the supercontinuum broadening is directed mainly through the anomalous dispersion region and towards the other normal dispersion region [44]. Such broadening mechanism can be explained by three-step process. First, initial broadening is due to SPM into a typical normal-dispersion continuum, where each pulse front broadens both spectrally and temporally. During the second step one of the pulse fronts reaches the ZDW and leaks into the anomalous dispersion region. In this process a soliton forms in the anomalous dispersion region as the anomalous dispersion compensates for the accumulated nonlinear phase. The soliton is repulsed spectrally from the continuum in the normal dispersion region because of the XPM between them. In the third step the dispersive waves (DWs) is originated in the other normal dispersion region, which occurs either as direct FWM with the soliton or as non-degenerate FWM between the soliton and the continuum in the other normal dispersion region (i.e. an XPM initiated process). During this step, the soliton repulsion is stopped and it recoils at the second ZDW [44].

4.4.5 MODULATION INSTABILITY AND FOUR-WAVE MIXING

In the study of nonlinear optics, FWM between narrowband optical waves is one of the most essential effects [45]. When two pump waves with identical frequency interact with each other, then the nonlinear interaction comprises the transformation of the pump into sidebands called Stokes (frequency downshifted) and anti-Stokes (frequency upshifted) waves. The frequency shift due to such nonlinear interaction is relative to the pump frequency. In such nonlinear interaction, a lot of key features of the FWM process are seen. Considering the un-depleted pump with power P_0, the evolution of the downshifted and upshifted sidebands is exponential with the gain amplitude (g) and given by the relation: $g = \sqrt{\left[\left(\gamma P_0 \right)^2 - \left(\frac{\Delta k}{2} \right)^2 \right]}$, where γ denotes the nonlinear coefficient, P_0 is the peak power of the pump wave, Δk represents the phase-mismatch and for the single-mode fiber it can be written as:

$$\Delta k = 2\gamma P_0 + 2\sum_{m=1}^{\infty}\left(\frac{\beta_{2m}}{(2m)!}\right)\Omega^{2m};$$ where β_{2m} depicts the even dispersion coefficients

of the fiber mode at pump wavelength, and Ω is the angular frequency shift from the pump frequency. β_{2m} corresponding to the Taylor series expansion of the propagation constant about the pump frequency. In the absence of initial seeding, the FWM corresponds to an instability in the propagating CW pump wave and the evolution from the noise of the sidebands symmetric about the pump frequency. The maximum growth rate is at the frequencies satisfying the phase-matching condition where $\Delta k = 0$, with the maximum gain amplitude, $g_{max} = \gamma P_0 = \dfrac{1}{L_{NL}}$. In the time-domain, this process leads to a temporal modulation, and the process is called as the modulation instability (MI) [5]. It is worthwhile to clear here that the MI and FWM can be considered as time-domain and frequency-domain descriptions of the same underlying physics [46]. But, the particular time-domain and frequency-domain depictions can be beneficial for explaining diverse aspects of the field evolution. Both the terms are frequently used in the literature. In the process of SCG, MI/FWM processes govern the mechanisms of spectral broadening in the case of input pulses of nanosecond-picosecond, or for the case of CW pumping. For the efficient broadening of the supercontinuum spectrum it is desirable to pump the fiber close to its ZDW. Though it is vital to note that the involvement of the higher-order dispersion terms to the phase-matching condition can produce substantial gain windows with the pump in either the normal or the anomalous dispersion regime [47, 48]. The analysis of the FWM has also been advantageous in describing the advent of the certain narrowband spectral landscapes in case of the excitation of femtosecond pulse due to the interaction of solitons and the DWs [49].

REFERENCES

[1] R. R. Alfano and S. L. Shapiro, "Emission in the region 4000 to 7000 Å via four-photon coupling in glass," *Phys. Rev. Lett.* 24, 584 (1970).

[2] C. Lin and R. H. Stolen, "New nanosecond continuum for excited-state spectroscopy," *Appl. Phys. Lett.* 28, 216 (1976).

[3] J. M. Dudley, G. Genty, and S. Coen, "Supercontinuum generation in photonic crystal fiber," *Rev. Mod. Phys.* 78(4), 1135–1184 (2006).

[4] G. Genty, S. Coen, and J. M. Dudley, "Fiber supercontinuum sources (Invited)," *J. Opt. Soc. Am. B* 24, 1771 (2007).

[5] J. M. Dudley and J. R. Taylor, *Supercontinuum Generation in Optical Fibers* (Cambridge University Press, 2010).

[6] C. R. Petersen, U. Moller, I. Kubat, B. Zhou, S. Dupont, J. Ramsay, T. Besson, S. Sujecki, N. Abdel-Moneim, Z. Tang, D. Furniss, A. Seddon, and O. Bang, "Mid-infrared supercontinuum covering the 1.4 – 13.3 μm molecular fingerprint region using ultra-high NA chalcogenide step-index fiber," *Nat. Photon.* 8, 830–834 (2014).

[7] T. S. Saini, A. Bailli, A. Kumar, R. Cherif, M. Zghal, and R. K. Sinha, "Design and analysis of equiangular spiral photonic crystal fiber for mid-infrared supercontinuum generation," *J. Modern Optics* 62(19), 1570–1576 (2015).

[8] AGN Chaitanya, T. S. Saini, A. Kumar, and R. K. Sinha, "Ultra broadband mid-IR supercontinuum generation in Ge11.5As24Se64.5 based chalcogenide graded-index photonic crystal fiber: design and analysis," *Appl. Opt.* 55(36), 10138 – 10145 (2016).

[9] T. S. Saini, A. Kumar, and R. K. Sinha, "Broadband mid-infrared supercontinuum spectra spanning 2 – 15 μm using As₂Se₃ chalcogenide glass triangular-core graded-index photonic crystal fiber," *IEEE/OSA J. Lightwave Technol.* 33(18), 3914 – 3920 (2015).

[10] T. Cheng, K. Nagasaka, T. H. Tuan, X. Xue, M. Matsumoto, H. Tezuka, T. Suzuki, and Y. Ohishi, "Mid-infrared supercontinuum generation spanning 2.0 – 15.1 μm in a chalcogenide step-index fiber," *Opt. Lett.* 41(9), 2117–2120 (2016).

[11] H. Takara, T. Ohara, T. Yamamoto, H. Masuda, M. Abe, H. Takahashi, and T. Morioka, "Field demonstration of over 1000-channel DWDM transmission with supercontinuum multi-carrier source," *Elect. Lett.* 41, 270–271 (2005).

[12] I. Hartl, X. D. Li, C. Chudoba, R. K. Ghanta, T. H. Ko, J. G. Fujimoto, J. K. Ranka, and R. S. Windeler, "Ultrahigh-resolution optical coherence tomography using continuum generation in an air-silica microstructure optical fiber," *Opt. Lett.* 26, 608–610 (2001).

[13] P. Hsiung, Y. Chen, T. H. Ko, J. G. Fujimoto, C. J. S. de Matos, S. V. Popov, J. R. Taylor, and V. P. Gapontsev, "Optical coherence tomography using a continuous-wave, high-power, Raman continuum light source," *Opt. Express* 12, 5287–5295 (2004).

[14] M. Yamanaka, H. Kawagoe, and N. Nishizawa, "High-power supercontinuum generation using high-repetition-rate ultrashort-pulse fiber laser for ultrahigh-resolution optical coherence tomography in 1600 nm spectral band," *Appl. Phys. Express* 9, 022701 (2016).

[15] C. Dunsby, P. M. P. Lanigan, J. McGinty, D. S. Elson, J. Requejo-Isidro, I. Munro, N. Galletly, F. McCann, B. Treanor, B. Onfelt, D. M. Davis, M. A. A. Neil, and P. M. W. French, "An electronically tunable ultrafast laser source applied to fluorescence imaging and fluorescence lifetime imaging microscopy," *J. Phys. D: Applied Physics* 37, 3296–3303 (2004).

[16] S. T. Sanders, "Wavelength-agile fiber laser using group-velocity dispersion of pulsed super-continua and application to broadband absorption spectroscopy," *Appl. Phys. B Lasers Opt.* 75, 799–802 (2002).

[17] M. Ere-Tassou, C. Przygodzki, E. Fertein, and H. Delbarre, "Femtosecond laser source for real-time atmospheric gas sensing in the UV-visible," *Opt. Commun.* 220, 215–221 (2003).

[18] J. Wegener, R. H. Wilson, and H. S. Tapp, "Mid-infrared spectroscopy for food analysis: recent new applications and relevant developments in sample presentation methods," *Trends Anal. Chem.* 18, 85–93 (1999).

[19] A.B. Seddon, "Mid-infrared (IR)-A hot topic: The potential for using mid-IR light for non-invasive early detection of skin cancer in vivo," *Phys. Status Solidi B.* 250(5), 1020–1027 (2013).

[20] A. B. Seddon, "A prospective for new mid-infrared medical endoscopy using chalcogenide glasses," *Int. J. Appl. Glass Sci.* 2, 177–191 (2011).

[21] A. Labruyere, A. Tonello, V. Couderc, G. Huss, and P. Leproux, "Compact supercontinuum sources and their biomedical applications," *Opt. Fiber Technol.* 18, 375–378 (2012).

[22] G. P. Agrawal, *Nonlinear Fiber Optics*, 5th ed., Elsevier Academic Press (Oxford, and Waltham, MA, 2013).

[23] W. Liu, L. Pang, X. Lin, R. Gao, and X. Song, "Observation of soliton fission in microstructured fiber," *Appl. Opt.* 51(34), 8095–8101 (2012).

[24] A. V. Husakov and J. Herrmann, "Supercontinuum generation of higher order solitons by fission in photonic crystal fibers," *Phys. Rev. Lett.* 87, 203901 (2001).

[25] G. Steinmeyer and J. S. Skibina, "Entering the mid-infrared," *Nat. Photon.* (news & views) 8(11), 814–815 (2014).

[26] K. Mori, H. Takara, S. Kawanishi, M. Saruwatari, and T. Morioka, Flatly broadened supercontinuum spectrum-generated in a dispersion decreasing fiber with convex dispersion profile. *Electron. Lett.* 33, 1806 (1997).

[27] T. S. Saini. N.P.T. Hoa, T. H. Tuan, X. Luo, T. Suzuki, and Y. Ohishi, "Tapered tellurite step-index optical fiber for coherent near-to-mid-IR supercontinuum generation: experiment and modeling," *Appl. Opt.* 58(2), 415–421 (2019).

[28] T. S. Saini. T. H. Tuan, T. Suzuki, and Y. Ohishi, "Coherent mid-IR supercontinuum generation using tapered chalcogenide step-index optical fiber: Experiment and modelling," *Sci. Rep.* 10, 2236 (2020).

[29] J. C. Travers, M. H. Fosz, and J. M. Dudley, "Nonlinear fiber optics overview," in J. M. Dudley and J. R. Taylor, eds., *Supercontinuum generation in optical fibers* Cambridge University Press (New York, 2010), pp. 32–51.

[30] M. H. Frosz, "Validation of input-noise model for simulations of supercontinuum generation and rogue waves," *Opt. Express* 18, 14778–14787 (2010).

[31] P. D. Drummond and J. F. Corney, "Quantum noise in optical fibers. I. Stochastic equations," *J. Opt. Soc. Am. B* 18, 139–152 (2001).

[32] K. Corwin, N. Newbury, J. Dudley, et al., "Fundamental noise limitations to supercontinuum generation in microstructure fiber," *Phys. Rev. Lett.* 90, 11 (2003).

[33] K. Saitoh, M. Koshiba, T. Hasegawa, and E. Sasaoka, "Chromatic dispersion control in photonic crystal fibers: Application to ultra-flattened dispersion," *Opt. Express* 11(8), 843–852 (2003).

[34] J. M. Hsu and B. L. Wang, "Tailoring of broadband dispersion-compensating photonic crystal fiber," *J. Mod. Opt.* 64(12), 1134–1145 (2016).

[35] S. Chandru and R. Anbazhagan, "Highly negative dispersion nanoscore spiral photonic crystal fiber for supercontinuum generation," *Nat. Acad. Sci. Lett.* 43, 153–155 (2020).

[36] S. Roy, S. K. Bhadra, K. Saitoh, M. Koshiba, and G. P. Agrawal, "Dynamics of Raman soliton during supercontinuum generation near the zero-dispersion wavelength of optical fibers," *Opt. Express* 19, 10443–10455 (2011).

[37] P. K. A. Wai, C. R. Menyuk, Y. C. Lee, and H. H. Chen, "Nonlinear pulse propagation in the neighborhood of the zero-dispersion wavelength of monomode optical fibers," *Opt. Lett.* 11, 464–466 (1986).

[38] N. Akhmediev, M. Karlsson, "Cherenkov radiation emitted by solitons in optical fibers," *Phys. Rev. A* 51, 2602–2607 (1995).

[39] J. P. Gordon, "Theory of the soliton self-frequency shift," *Opt. Lett.* 11, 662–664 (1986).

[40] A. A. Voronin and A. M. Zheltikov, "Soliton self-frequency shift decelerated by self-steepening," *Opt. Lett.* 33(15), 1723–1725 (2008).

[41] N. Nishizawa and T. Goto, "Characteristics of pulse trapping by use of ultrashort soliton pulses in optical fibers across the zero-dispersion wavelength," *Opt. Express* 10, 1151–1159 (2002).

[42] G. Genty, M. Lehtonen, and H. Ludvigsen, "Effect of cross-phase modulation on supercontinuum generated in microstructured fibers with sub-30 fs pulses," *Opt. Express* 12, 4614–4624 (2004).

[43] A. V. Gorbach and D. V. Skryabin, "Light trapping in gravity-like potentials and expansion of supercontinuum spectra in photonic-crystal fibers," *Nat. Photon.* 1, 653–657 (2007).

[44] S. Christensen, D. S. Shreesha Rao, O. Bang, and M. Bache, "Directional supercontinuum generation: the role of the soliton," *J. Opt. Soc. Am. B* 36(2), A131 (2019).

[45] R. H. Stolen and J. E. Bjorkholm, "Parametric amplification and frequency conversion in optical fibers," *IEEE J. Quantum Electron.* QE-18, 1062–1072 (1982).

[46] R. Stolen, J. Gordon, W. Tomlinson, and H. Haus, "Raman response function of silica-core fibers," *J. Opt. Soc. Am. B* 6, 6, 1159–1166 (1989).

[47] J. D. Harvey, R. Leonhardt, S. Coen, et al., "Scalar modulation instability in the normal dispersion regime by use of a photonic crystal fiber," *Opt. Lett.* 28, 2225–2227 (2003).

[48] W. H. Reeves, D. V. Skryabin, F. Biancalana, et al., "Transformation and control of ultra-short pulses in dispersion-engineered photonic crystal fibers,"' *Nature.* 424, 511–515 (2003).

[49] D.V. Skryabin and A. V. Yulin, "Theory of generation of new frequencies by mixing of solitons and dispersive waves in optical fibers," *Phys. Rev. E* 72, 016619/1–10 (2005).

5 Photonic Crystal Fiber Technologies

5.1 INTRODUCTION

In this chapter, the novel built-in functionalities achievable in optical fiber technology that does more important things than the transmission of light are discussed. Dropping the complexity of the optical systems emphasize the need to upsurge the number of optical functions in as few modules as possible, making the system more compact and simple to operate. Optical functionalities such as light generation, broadband transmission, amplification, frequency conversion, wavelength filtering, and interfacing to the outside environment typically needed separate, individual optical components to be chained together to comprise a system. However, with the initiation of photonic crystal fiber, many of these functionalities can be pooled into multifunctional modules paving the way for simpler, turn-key systems. Some of the state-of-the-art PCF technologies to obtain multi-functionalities in the PCF structures are discussed in the next sections.

5.2 NANOTECHNOLOGIES ON PHOTONIC CRYSTAL FIBER PLATFORM

Based on the principle of light guidance, there are two broad types of PCFs that exist: one type of PCF allows to guide light by photonic bandgap effect, and one allows to guide light by the principle called modified total internal reflection (M-TIR), in which the light is confined in the core with a higher refractive index obtained by introducing micro-size air holes in the cladding region [1, 2]. PCFs have exceptional characteristics that are not possessed by the standard optical fibers, for example, broadband dispersion engineering, endless single-mode operation, high-nonlinearities, high-numerical aperture for high-power cladding-pumped fiber lasers, and high birefringence [3–6]. PCFs have additional exceptional properties: they possess micro-holes surrounding the core region. The advantage of these air holes is that they can be filled with tunable liquid materials, hence opening up a new opportunity to fabricate all-in-fiber tunable photonic devices. All-in-fiber photonic devices are very attractive as they are compact and can easily be coupled to the fibers, therefore absconding alignment issues. Furthermore, being based on the fiber platform, they reveal circular symmetry and, therefore, polarization sensitivity, such as polarization mode dispersion (PMD) and polarization-dependent loss (PDL) can be reduced compared to the planar-based components. Additionally, tunability is an attractive property since tunable filters are, for example, needed in the fabrication of tunable fiber lasers, in reconfigurable optical networks, and in sensing.

DOI: 10.1201/9781003502401-5

The fibers in which different matrix materials can be used were developed by the stack and draw technique. Using the stack and draw technique, it is easy to combine different materials within the structure provided they can be drawn together. This technique has been used to produce hybrid guiding structures, in which some of the air holes are filled with doped glass, such as all-solid bandgap structures, lasers based on microstructured optical fiber [7–9]. In another way, liquids can be used to infiltrate the holes of the photonic crystal fibers after the drawing. Filling the air holes of the microstructure with a higher index material can alter a regular hexagonal stacked fiber from an index guiding to the photonic bandgap guiding structure [10, 11]. This approach of filling the air holes permits the properties of the bandgap fiber to develop switchable photonic devices [12].

It is possible to transmute the PCF that guided light based on the M-TIR into a photonic bandgap (PBG) waveguide by filling high-index the liquid in its air holes in the cladding region [13]. Also, the location of the bandgaps could be tuned by tuning the temperature of the high-index liquid filled in air holes. Polymers can also be infiltrated in a special type of PCF called "grapefruit"-shaped microstructured optical fiber [14]. By using this polymer filling technique, tunable gratings [14, 15] and tunable birefringence [16] are demonstrated. In these reports, the tunability was realized by altering the temperature of the infiltrated polymer. Extra development in this field is made by the feasibility of filling liquid crystals in PCFs. Liquid crystals are organic materials that exhibit very high thermo-optic and electro-optic effects because of the high dielectric anisotropy and high birefringence. Liquid crystals have a response time very short in comparison to the high-index liquids and polymers. Additionally, anisotropy in the liquid crystals gives extra degrees of freedom than the conventional isotropic liquids. The various photonic devices based on the PCFs filled with liquid crystal have been demonstrated in the literature. In such liquid crystal-filled PCF devices, the electrical [17–19], thermal [20, 21], and the optical tunability [22] has been demonstrated. The waveguiding mechanism of such structures of liquid crystal-filled PCFs has been studied in the literature [23–26].

The properties of the photonic crystal fibers can be adjusted by successfully filling or coating the air holes using metal nanoparticles/layers. This is an interesting feature of the photonic crystal fibers and useful for the important application for which selective filling or coating is required. Recently, an impressive experiment demonstrated that various materials including semiconductors could be incorporated into photonic crystal fibers using high-pressure microfluidic chemical deposition [27]. The integration of emerging PCF technology with surface-enhanced Raman scattering (SERS) can be helpful to develop the prevailing SERS/PCF platform for ultra-trace biological and chemical sensing and molecular fingerprinting. The evidence of such a platform rests on the strong mode-field overlap with an analyte over the whole length of the PCF and the advantage offered by SERS. The main thing for the understanding of SERS/PCF is the controlled immobilization of SERS-active Au or Ag nanoparticles inside the longitudinally aligned air channels in the silica PCF with air cladding microstructure optimized for sensing applications.

The combination of photonics and plasmonics is an evolving field that would advantage from enhancements in coating techniques. This area is attracting increasing interest in photonic crystal fiber research. The surface plasmon resonance sensor

based on microstructured optical fiber with optimized microfluidics has been reported
[28]. The plasmons can be excited on the inner surface of the large metalized channels containing analyte by the fundamental mode of a single-mode microstructured
optical fiber. By introducing air-filled microstructure into the fiber core the phase
matching between core mode and plasmon can be enforced. This allows the tuning of
the modal refractive index and its matching with that of a plasmon. Figure 5.1 shows
the transverse cross-sectional structure of a microstructured optical fiber-based surface plasmon resonance (SPR) sensor. One air holes are at the centre of the core of
the microstructured fiber structure. The core of the fiber is surrounded by two rings
of air holes. The holes in the first ring and at the centre of the core are filled with air
$n_{air} = 1.0$. The holes in the second ring are metalized and the size of the holes is larger
than the air hole size in the first ring. The holes in the second ring are filled with an
analyte (water) $n_a = 1.33$. The diameters of the holes in the first and second layers are
$d_1 = 0.6L$ and $d_2 = 0.8L$, respectively. L is the pitch of the air hole arranged in the
hexagonal lattice. The pitch of the underlying hexagonal lattice of the air hole is L =
2mm. As mentioned above, the core of a microstructured optical fiber features a central air hole of diameter $d_c = 0.45L$. The region of having a small air hole at the centre
of the core of the microstructured optical fiber is to allow phase matching with the
plasmon to lower the refractive index of the core-guided mode. The air holes in the
second ring are filled with the analyte and metalized for the plasmon excitation. The
light guides in the higher refractive index of the core region. Also, the air holes in the
first ring control the coupling strengths between the core mode and the plasmon
mode. The effective index of the fundamental mode can be tuned by changing the
size of the central air hole in the core region. The region of the first ring of the air
holes works as a low-index cladding enabling guidance of light in the core of the

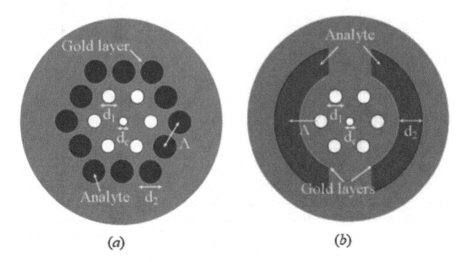

(a)   (b)

FIGURE 5.1   (a) The transverse cross-sectional view of a microstructured optical fiber-based
SPR sensor.

Adapted from [28].

microstructured optical fiber. The size of the air holes in the first ring influences strongly the coupling strength between the plasmon mode and the core mode. It has been found that a larger hole size results in weaker coupling, thus longer the size of the sensors. The most important thing is the holes in the second ring are metalized with a gold layer of a thickness of 40nm and feature larger diameters to enable smooth flow of an analyte.

Figure 5.2 illustrates the calculated loss of the spectrum of the mode guided in the core of the designed microstructured optical fiber at different refractive indices of the analytes. In the figure, three-loss peaks corresponding to the excitation of the various plasmonic modes in the metalized holes are revealed. The black solid line corresponds to $n_a = 1.33$, the blue dotted line corresponds to $n_a = 1.34$. For the comparison, a red-dashed line shows the confinement loss of the mode confined in the core in the absence of the metal coating. As shown in the figure, the resonant frequency of the first plasmon peak is around 560 nm. It is to be noted here that the shape of the metalized surface modifies the plasmonic excitation spectrum. The planar metalized surface offers only one plasmonic peak, while the cylindrical metal layer can support several different plasmonic peaks [29–32].

The calculated loss spectra for the first plasmonic peak at various thicknesses of the gold coating i.e. 30nm, 40nm, and 50nm are shown in Figure 5.3. Figure 5.3 illustrates the changes in positions of a plasmonic peak with the thickness of the gold

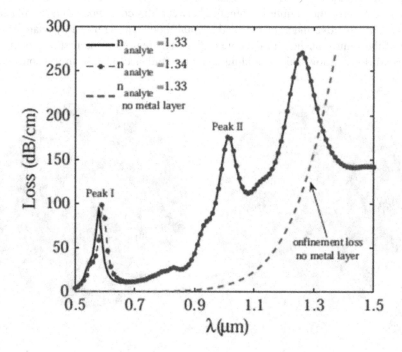

FIGURE 5.2   The illustration of calculated loss of the spectrum of the designed microstructured optical fiber core-guided mode.

Adapted from [28].

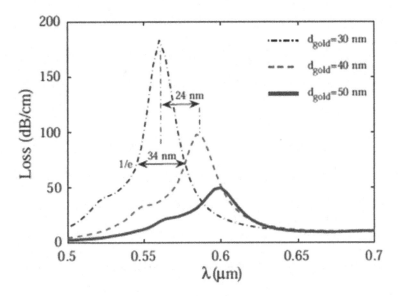

FIGURE 5.3 Calculated loss spectra of the first plasmonic peak for 30nm, 40nm and 50nm thicknesses of a gold coating.

Adapted from [28].

layer varying to 30nm, 40nm, and 50nm. As the thickness of the gold layer increases the modal loss at the resonance decreases. At the same instant, there is a shift in the loss peak towards the longer wavelengths. For instance, the plasmonic peak corresponding to the thickness of 30nm gold film shifts 24 nm towards the longer wavelengths when the layer thickness is altered to 40nm. The shift in the loss peak position because of the nanometer variations in the thickness of the metal layer can be easily noticed. This transduction mechanism is very important to study metal nanoparticle binding events to the walls of the metalized holes. Such a type of sensor operation is attractive to monitor the concentration of the nanoparticles used as carriers of the photosensitive drugs in photodynamic cancer therapy [33].

The traditional mode of operation of the proposed microstructured optical fiber-based sensor would be the detection of changes in the refractive index of the analyte in the quick vicinity of the metal surface. The sensitivity analysis of the designed microstructured optical fiber-based SPR sensor for various thicknesses of the gold coating i.e. 30nm, 40nm, 50nm and 65nm are shown in Figure 5.4. As illustrated in Figure 5.4, the sensitivity of the designed microstructured optical fiber depends weakly on the thickness of the gold layer. For the thinner metal films, the peak of the sensitivity curve shifts towards the shorter wavelengths. Also, the change in the refractive index of analyte by 10^{-4} results in at least a 1% change in the transmitted intensity. This is a fair comparison to what is obtained in conventional fiber-based SPR sensors. It is very clear to note here that due to the very high loss of the designed microstructured optical fiber-based plasmonic sensor, its length is limited to the several centimetres. Therefore, such a microstructured optical fiber metal-coated sensor

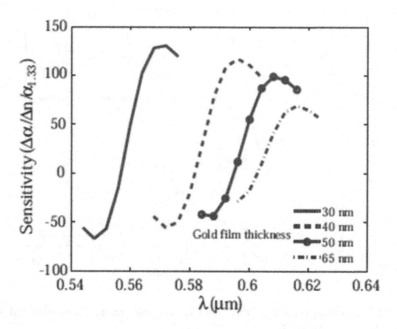

FIGURE 5.4 The sensitivity analysis of the microstructured optical fiber-based SPR sensor for the various thicknesses of the gold coating layer.

Adapted from [28].

can be considered as the integrated photonic elements rather than the conventional fiber-based sensors.

For various applications including index guidance in liquids, birefringent fiber, production of non-uniform tapers in which the photonic structure varies longitudinally, and the surface plasmon effects, the selective filling or closing of the air holes at the fiber stage is gaining a lot of attention [34–39]. The selective filling and blocking of the air holes at the fiber stage have gained significant attention. A selective coating in the air holes of the microstructured optical fibers with the metal has been performed using a two-stage drawing process, and an in-fiber absorptive polarizer is demonstrated [40]. In this work, first of all, a primary preform (with a diameter of 70 mm made by polymethylmethacrylate rods) was drilled with the desired hole pattern using the computer-controlled mill and drawn to an intermediate 'cane' or perform with 1 cm diameter which is then drawn to the fiber. The image of the fabricated preform is shown in Figure 5.5.

The experimental setup employed for the silver deposition is illustrated in Figure 5.6. To selectively block the air holes, the intermediate preform was glued using industry-grade silicon prior to solution coating. Then the fiber end of the intermediate preform is connected to the syringe, permitting suction of the reaction mixture through the unblocked holes. The reaction mixture is maintained at 5 °C to prevent excess deposition of silver and to avoid the blockage at the fiber preform end. The residual reaction mixture can be removed by suction of water rather than air. The selective coating was performed on a five-fold symmetric microstructured fiber

FIGURE 5.5 The image of the fabricated fiber preform or 'cane'.

Adapted from [40].

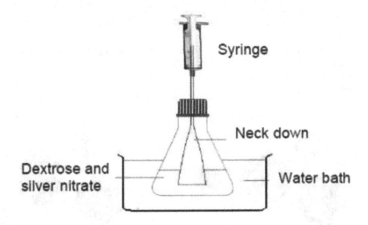

FIGURE 5.6 The setup employed for the silver deposition.

Adapted from [40].

having relatively large holes in the first ring. Figure 5.7 illustrates the structure rep-
resenting the coated air holes. The diameter of the small air hole in the second ring
in which selective silver coating was successfully performed is 5 μm. The coatings
are granular and form a well-defined, adhering layer on the inner walls of the fiber
structure. The coating length of up to 40 cm was coated by this technique. The
x-polarized and y-polarized fundamental modes of the structure are shown in
Figure 5.8. The vector plots of the transverse components of the electric field for the
fundamental x-polarized fundamental mode, with 7.24dB/m losses (left) and
y-polarized fundamental mode, with 1.26dB/m losses (right). The red circles represent the holes coated with silver metal. The y-polarized modes are ascribed to the fact
that these modes have an electric field mainly parallel to the large silver-coated hole,
and therefore cannot couple simply to surface plasmonic modes, while the x-polarized
modes have an electric field mainly orthogonal to the silver surface and couple more

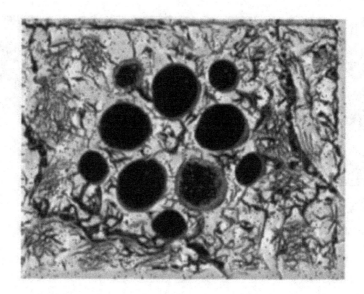

FIGURE 5.7 The microscopic image of microstructured optical fiber with selectively coating in two holes.

Adapted from [40].

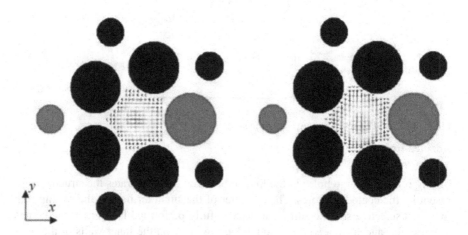

FIGURE 5.8 The electric field pattern of the x-polarized and y-polarized fundamental modes distribution in the core of the selectively coated microstructured optical fiber structure.

Adapted from [40].

actively to the surface plasmons. This phenomenon appears to be only weakly resonant, with negligible wavelength dependence. While the SPR generally have strong wavelength dependence, modelling directs the occurrence of multiple resonances that slur out to produce an effect that is impartially constant with the wavelength.

The polarizing properties of the coated fiber of the length of 2 cm were character-ized by the experimental setup shown in Figure 5.9. The experiment setup was con-sidered to have no movable components in front of the fiber to avoid any alteration in the beam position which could change the excitation mode distribution. A circularly polarized light beam from a HeNe laser (632.8nm line) was launched into the coated microstructured optical fiber by employing a quarter-wave plate. At the detection end, a pinhole was used to segregate the core light exiting the microstructured optical fiber. The intensity of core light at distinct polarizations over 360 degrees was sam-pled for every 10 degrees using a rotation of the polarizer (P_2). Finally, the optical power was measured using a photodiode-2 without the sample fiber used as a control. To confirm the attenuation measurements were not affected by the drift, both the signal and control measurements were normalized to the total power launched into the system, which was recorded by photodiode-1. The length of the fiber employed in the experiment was 2 cm and well-maintained in a straight line, bend losses or torsion effects are supposed to be insignificant. The transmission outcomes as a func-tion of the rotation of P_2 are exposed in Figure 5.10. The coated fiber region is detected to reduce the intensity of core light in one polarization by around 50%. The subsequent polarization behaviour of the fiber reflects both, the differential attenua-tion of the two polarization states and possibly their different launch conditions. With the optimized launching conditions, light is primarily coupled to the two polariza-tions of the fundamental mode, with the $LP_{1,2}$ being only weakly excited.

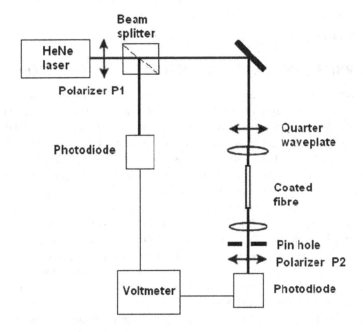

FIGURE 5.9 The polarization measurement setup.

Adapted from [40].

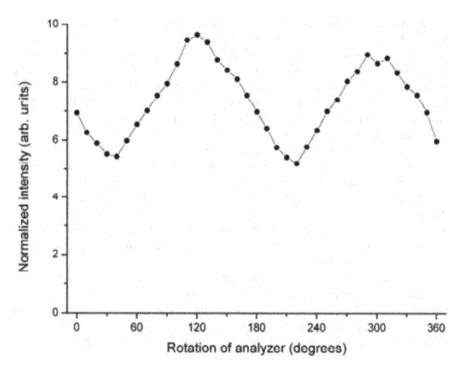

FIGURE 5.10 Transmission intensity through the coated fiber, as a function of the rotation of the polarizer P2.

Adapted from [40].

5.3   ORBITAL ANGULAR MOMENTUM IN OPTICAL FIBERS

To increase the capacity of optical fiber links, orbital angular momentum (OAM) of the light is an encouraging resource for the exploitation of the spatial dimension of the light. Traditionally, it is realized by multiplexing the wavelength, spatial degree of freedom, and the polarization of light to increase the data channels for a parallel transformation. The OAM of the light wave was documented by Les Allen in 1992 [41]. A detailed summary of how OAM enables efficient mode multiplexing for optical communications, with emphasis on the design of OAM fibers is reviewed in [42]. The OAM of light has been deliberated as a capable degree of freedom for the multiplexing data in free space [43–45]. The orthogonality of OAM with other multiplexing domains has been investigated and the free-space data link that uniquely combines OAM, polarization, and the wavelength division multiplexing is presented. It has been demonstrated that the multiplexing/demultiplexing of 1008 data channels carried on two polarizations, 12 OAM beams, and 42 wavelengths can be carried out. [43].

A light beam carrying OAM possesses a helical wavefront, which is described by the relation: $\exp(il\phi)$, where ϕ represents the azimuthal angle, and the topological charge l is an unbounded integer. Superior to spin angular momentum (i.e. circular polarization) with two states, OAM could offer unrestricted channels for data transmission. Due to this exceptional property, OAM multiplexing has been extensively applied to realize high-capacity communication in optical fibers and free space

[46–50]. Nevertheless, optical communication in free space using OAM multiplexing unavoidably suffers from the issue of multiple scattering from ambient microparticles in the atmosphere [51–53]. Due to this, it is conceivable that the wavefronts of the OAM modes get scrambled and destroy the orthogonality between the OAM channels [54]. The deteriorated orthogonality will upsurge the crosstalk when the information carried in different OAM channels is demultiplexed [48, 55]. Subsequently, it is stimulating to realize OAM-based communication through the scattering media. In addition to scattering media, because of the mode coupling and dispersion, optical transmission through multimode systems including waveguides and the fibers also suffers from the same problems. Recently, a scattering-matrix-assisted retrieval technique (SMART) has been proposed to precisely extract encoded OAM states from the multiply scattered light [56]. This SMART first uses a speckle-correlation scattering matrix to recuperate the optical field of a data-carrying vortex beam with the superposition of the OAM states and then demultiplexes each OAM channel with the mode decomposition method.

To corroborate the SMART platform experimentally, an optical data transmission link based on a digital micromirror device (DMD) built and shown in Figure 5.11.

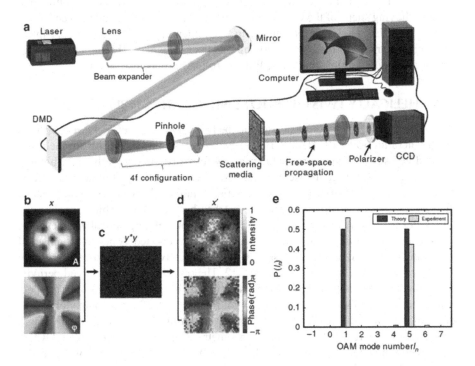

FIGURE 5.11 The experimental setup and characterization of the SMART platform. (a) The experimental setup of the SMART platform. (b–d) Field retrieval of a known incident field. For a given LG superposition field (x; (b)), a raw intensity speckle with a single shot (y*y; (c)) is recorded. The retrieved field (x'; (d)) is achieved by using the SMART. The symbols A and φ denote the amplitude and phase of the fields, respectively. (e) A comparison between the measured OAM spectrum by the SMART and the theoretical spectrum.

Adapted from [56].

DMD is employed to encode parallel OAM channels and realize reference-free calibration in a multiple-scattering environment. Due to the ability of the complex field modulation and a high revive rate, a single DMD enables high-fidelity generation of OAM fields and quick switching amongst them. An optical diffuser (DG10–220, Thorlabs, Inc.) is used to imitate an optically scattering environment and inserted in the transmission path. To achieve rapid reference-free calibration in the setup a technique developed from the parallel wavefront optimization method can be used to attain quick reference-free calibration in the same setup [57]. By moderating the wavefront of the incident light, the TM, an assembly of diffused fields for all input modes, can be derived from the transmitted speckle patterns with the calibration algorithm. In the technology of OAM multiplexing, the orthogonality between OAM channels guarantees efficient (de)multiplexing. The binary data supported in multiplexed OAM states can be encoded into a single laser beam for the purpose of communication.

The first designed fiber, especially for the transmission of OAM modes, is the vortex fiber [58]. The vortex fiber is characterized by a central core surrounded by a ring core, for the transmission of the first OAM mode. The centre core is for the transmission of the fundamental mode. In this fiber, the idea was to design a fiber that mimics the ring shape of the OAM modes and with a large separation between $TE_{0,1}$, $TM_{0,1}$, and $HE_{2,1}$ modes, to prevent them from coupling into $LP_{1,1}$ mode. A lot of experiments are performed by employing this fiber design, combining OAM modes with wavelength multiplexing and QPSK or 16-QAM transmission [46].

5.4 HELICALLY TWISTED PHOTONIC CRYSTAL FIBERS

The experimental and theoretical work on the helically twisted PCFs is well reported in the literature [59, 60]. Helical Bloch theory is also introduced for the helically twisted PCFs [60]. The theory includes a new formalism based on the tight-binding approximation. It is used to discover and describe a multiplicity of unfamiliar effects that appear in a range of diverse twisted PCFs, including fibers with a single core and fibers with N cores arranged in a ring around the fiber axis. A novel birefringence that originates the propagation constants of left- and right-spinning optical vortices to be non-degenerate for the same order of OAM has been reported. A new mechanism of light guidance based on the coreless PCF that has been twisted its axis during the drawing process is shown in Figure 5.12. The resulting helically curved periodic 'space' generates a topological channel within which light can be strongly trapped, with a confinement strength that scales with the twist rate. Such a completely novel form of the waveguide depends on the quadratic upsurge in the optical path length with a radius that results from its helical geometry. This makes a potential well where light is confined by the photonic bandgap (PBG) effects at the bottom of the fundamental passband. The guidance mechanism is extremely unusual because cleaving the twisted fiber and examining its cross-section exposes no core structure at which light could be trapped. The twisted multicore PCFs have been revealed to support helical Bloch waves and to expose OAM birefringence [61]. There have also been many theoretical studies, for example, of spin-orbit coupling [62, 63] and Bloch dynamics in helical coupled waveguide arrays [64]. The solid-core twisted PCF

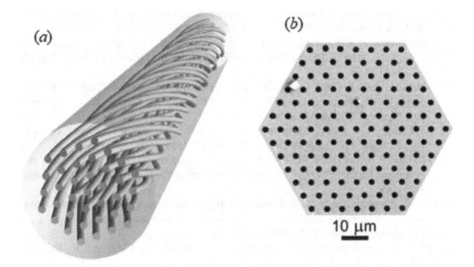

FIGURE 5.12 The geometry of the twisted coreless PCFs. (a) The schematic of a twisted coreless PCF. The axis of rotation coincides with the hollow channel in the centre. (b) The scanning electron microscopic image of the transverse cross-section of the microstructure.

Adapted from [62].

demonstrates a series of transmission dips at twist-tunable wavelengths, which offers prospective applications of twisted PCF in sensing and filtering. The hollow-core single-ring twisted PCF allows improved elimination of the unwanted higher-order guided modes. It is expected that many of these effects and phenomena will move into real-world applications in the near future.

REFERENCES

[1] J. Knight, J. Broeng, T. A. Birks, and P. Russell, "Photonic band gap guidance in optical fibers," *Science* 282(5393), 1476–1478 (1998).

[2] P. Russell, "Photonic crystal fibers," *Science* 299(5605), 358–362 (2003).

[3] J. Knight, T. Birks, P. S. J. Russell, and D. Atkin, "All-silica single mode optical fiber with photonic crystal cladding," *Opt. Lett.* 21(19), 1547–1549 (1996).

[4] T. Birks, J. Knight, and P. Russell, "Endlessly single-mode photonic crystal fiber," *Opt. Lett.* 22(13), 961–963 (1997).

[5] J. Ranka, R. Windeler, and A. Stentz, "Visible continuum generation in air-silica micro-structure optical fibers with anomalous dispersion at 800 nm," *Opt. Lett.* 25(1), 25–27 (2000).

[6] J. Limpert, T. Schreiber, S. Nolte, H. Zellmer, A. Tunnermann, R. Iliew, F. Lederer, J. Broeng, G. Vienne, A. Petersson, and C. Jakobsen, "High-power air-clad large-mode-area photonic crystal fiber laser," *Opt. Express* 11(7), 818–823 (2003).

[7] W. J. Wadsworth, J. C. Knight, W. H. Reeves, and P. St. J. Russell, "Yb³⁺-doped photonic crystal fiber laser," *Electron. Lett.* 36, 1452–1453 (2000).

[8] A. Argyros, T. Birks, S. Leon-Saval, C. M. Cordeiro, F. Luan, and P. S. J. Russell, "Photonic bandgap with an index step of one percent," *Opt. Express* 13, 309–314 (2005).

[9] A. S. Cerqueira Jr., F. Luan, C. M. B. Cordeiro, A. K. George, and J. C. Knight, "Hybrid photonic crystal fiber," *Opt. Express* 14, 926–931 (2006).

[10] C. Kerbage, P. Steinvurzel, P. Reyes, P. S. Westbrook, R. S. Windeler, A. Hale, and B. J. Eggleton, "Highly tunable birefringent microstructured optical fiber," *Opt. Lett.* 27, 842–844 (2002).

[11] N. Litchinitser, S. Dunn, P. Steinvurzel, B. Eggleton, T. White, R. McPhedran, and C. M. de Sterke, "Application of an ARROW model for designing tunable photonic devices," *Opt. Express* 12, 1540–1550 (2004).

[12] T. Larsen, A. Bjarklev, D. Hermann, and J. Broeng, "Optical devices based on liquid crystal photonic bandgap fibers," *Opt. Express* 11, 2589–2596 (2003).

[13] R. Bise, R. Windeler, K. Kranz, C. Kerbage, B. Eggleton, and D. Trevor, "Tunable photonic band gap fiber," *Optical Fiber Communication Conference OFC 2002*, pp. 466–468 (2002).

[14] P. Westbrook, B. Eggleton, R. Windeler, A. Hale, T. Strasser, and G. Burdge, "Cladding-mode resonances in hybrid polymer-silica microstructured optical fiber gratings," *IEEE Photon. Technol. Lett.* 12(5), 495–497 (2000).

[15] B. Eggleton, C. Kerbage, P. Westbrook, R. Windeler, A. Hale, and B. Eggleton, "Microstructured optical fiber devices," *Opt. Express* 9(13), 698–713 (2001).

[16] C. Kerbage, B. Eggleton, and B. Eggleton, "Numerical analysis and experimental design of tunable birefringence in microstructured optical fiber," *Opt. Express* 10(5), 246–255 (2002).

[17] M. Haakestad, T. Alkeskjold, M. Nielsen, L. Scolari, J. Riishede, H. Engan, and A. Bjarklev, "Electrically tunable photonic bandgap guidance in a liquid-crystal-filled photonic crystal fiber," *IEEE Photon. Technol. Lett.* 17(4), 819–821 (2005).

[18] F. Du, Y.-Q. Lu, and S.-T. Wu, "Electrically tunable liquid-crystal photonic crystal fiber," *Appl. Phys. Lett.* 85(12), 2181–2183 (2004).

[19] L. Scolari, T. T. Alkeskjold, J. Riishede, A. Bjarklev, D. S. Hermann, Anawati, M. D. Nielsen, and P. Bassi, "Continuously tunable devices based on electrical control of dual-frequency liquid crystal filled photonic bandgap fibers," *Opt. Express* 13(19), 7483–7496 (2005).

[20] T. Larsen, A. Bjarklev, and D. Hermann, "Optical devices based on liquid crystal photonic bandgap fibers," *Opt. Express* 11(20) (2003).

[21] T. R. Wolinski, K. Szaniawska, S. Ertman, P. Lesiak, A. W. Domanski, R. Dabrowski, E. Nowinowski-Kruszelnicki, and J. Wojcik, "Influence of temperature and electrical fields on propagation properties of photonic liquid-crystal fibers," *Measurem. Sci. Technol.* 17(5), 985–991 (2006).

[22] T. Alkeskjold, J. Laegsgaard, A. Bjarklev, D. Hermann, Anawati, J. Broeng, J. Li, and S.-T. Wu, "All-optical modulation in dyedoped nematic liquid crystal photonic bandgap fibers," *Opt. Express* 12(24) (2004).

[23] J. Lægsgaard, "Gap formation and guided modes in photonic bandgap fibers with high-index rods," *J. Opt. A Pure Appl. Opt.* 6(8), 798–804 (2004).

[24] D. Zografopoulos, E. Kriezis, and T. Tsiboukis, "Tunable highly birefringent bandgap-guiding liquid-crystal microstructured fibers," *J. Lightwave Technol.* 24(9), 3427–3432 (2006).

[25] J. Lægsgaard and T. Alkeskjold, "Designing a photonic bandgap fiber for thermo-optic switching," *J. Opt. Soc. Am. B Opt. Phys.* 23(5), 951–957 (2006).

[26] J. Sun, C. Chan, and N. Ni, "Analysis of photonic crystal fibers infiltrated with nematic liquid crystal," *Opt. Commun.* 278(1), 66–70 (2007).

[27] P. J. A. Sazio, A Amezcua-Correa, C. E. Finlayson, J. R. Hayes, T. J. Scheidemantel, N. F. Baril, B. R. Jackson, D-J Won, F. Zhang, E. R. Margine, V. Gopalan, V. H. Crespi, and J. V. Badding, "Microstructured optical fibers as high-pressure microfluidic reactors," *Science* 311, 1583–1586 (2006).

[28] A. Hassani and M. Skorobogatiy, "Design of the microstructured optical fiber-based surface plasmon resonance sensors with enhanced microfluidics," *Opt. Express* 14, 11616–11621 (2006); *Josa B* (2007).

[29] S. J. Al-Bader and M. Imtaar, "Optical fiber hybrid-surface plasmon polaritons," *J. Opt. Soc. Am. B* 10, 83 (1993).

[30] A. Diez, M. V. Andres, and J. L. Cruz, "In-line fiber-optic sensors based on the excitation of surface plasma modes in metal-coated tapered fibers," *Sens. Actuators B* 73, 95 (2001).

[31] D. Monzon-Hernandez, J. Villatoro, D. Talavera, and D. Luna-Moreno, "Optical-fiber surface-plasmon resonance sensor with multiple resonance peaks," *Appl. Opt.* 43, 1216 (2004).

[32] D. Monzon-Hernandez and J. Villatoro, "High-resolution refractive index sensing by means of a multiple-peak surface plasmon resonance optical fiber sensor," *Sens. Actuators B* 115, 227 (2006).

[33] L.O. Cinteza, T. Ohulchanskyy, Y. Sahoo, E. J. Bergey, R. K. Pandey, and P. N. Prasad, "Diacyllipid micelle-based nanocarrier for magnetically guided delivery of drugs in photodynamic therapy," *Mol. Pharm.* 3, 415 (2006).

[34] G. Vienne, M. Yan, T. Luo, T. K. Liang, P. Ho, and C. Lin, "Liquid core fibers based on hollow core microstructured fibers," in *Proceedings of IEE conference on lasers and electrooptics/Pacific Rim* (Institute of Electrical and Electronics Engineers40Tokyo, 2005), 551–552 (2005).

[35] F. M. Cox, A. Argyros, and M. C. J. Large, "Liquid-filled hollow core microstructured polymer optical fiber," *Opt. Express* 14, 4135–4140 (2006).

[36] A. Witkowska, K. Lai, S. G. Leon-Saval, W. J. Wadsworth, and T. A. Birks, "All-fiber anamorphic coreshape transitions," *Opt. Lett.* 31, 2672–2674 (2006).

[37] B. T. Kuhlmey, K. Pathmanandavel, and R. C. McPhedran, "Multipole analysis of photonic crystal fibers with coated inclusions," *Opt. Express* 14, 10851–10864 (2006).

[38] A. Hassani and M. Skorobogatiy, "Design of microstructured optical fiber-based surface plasmon resonance sensors with enhanced microfuidics", *Opt. Express* 14, 11616–11621 (2006).

[39] A. Hassani and M. Skorobogatiy, "Design criteria for microstructured-optical-fiber-based surface-plasmon-resonance sensors," *J. Opt. Soc. Am. B* 24, 1423–1429 (2007).

[40] X. Zhang, R. Wang, F. M. Cox, B.T. Kuhlmey, and M. C. J. Large, "Selective coating of holes in microstructured optical fiber and its application to in-fiber absorptive polarizers," *Opt. Express* 15, 16270–16278 (2007).

[41] L. Allen, et al., "Orbital angular momentum of light and the transformation of Laguerre-Gaussian laser modes," *Phys. Rev. A* 45, 8185–8189 (1992).

[42] C. Brunet and L. A. Rusch, "Optical fibers for the transmission of orbital angular momentum modes," *Opt. Fiber Technol.* 35, 2–7 (2017).

[43] H. Huang, et al., "100 Tbit/s free-space data link enabled by three-dimensional multiplexing of orbital angular momentum, polarization, and wavelength," *Opt. Lett.* 39, 197–200 (2014).

[44] A. E. Willner, et al., "Optical communications using orbital angular momentum beams," *Adv. Opt. Photon.* 7, 66–106 (2015).

[45] A. E. Willner, et al., "Recent advances in high-capacity free-space optical and radio-frequency communications using orbital angular momentum multiplexing," *Philos. Trans. R. Soc. A Math. Phys. Eng. Sci.* 375, 20150439 (2017).

[46] N. Bozinovic, et al., "Terabit-scale orbital angular momentum mode division multiplexing in fibers," *Science* 340, 1545–1548 (2013).

[47] L. Zhu, et al., "Orbital angular momentum mode multiplexed transmission in heterogeneous few-mode and multi-mode fiber network," *Opt. Lett.* 43, 1894–1897 (2018).

[48] J. Wang, et al., "Terabit free-space data transmission employing orbital angular momentum multiplexing," *Nat. Photon.* 6, 488–496 (2012).

[49] M. Krenn, et al., "Twisted light transmission over 143 km," *Proc. Natl Acad. Sci. USA* 113, 13648–13653 (2016).

[50] M. P. J. Lavery, et al., "Free-space propagation of high-dimensional structured optical fields in an urban environment," *Sci. Adv.* 3, e1700552 (2017).

[51] N. B. Zhao, et al., "Capacity limits of spatially multiplexed free-space communication," *Nat. Photon.* 9, 822–826 (2015).

[52] C. Z. Shi, et al., "High-speed acoustic communication by multiplexing orbital angular momentum," *Proc. Natl Acad. Sci. USA* 114, 7250–7253 (2017).

[53] W. B. Wang, et al., Deep transmission of Laguerre–Gaussian vortex beams through turbid scattering media," *Opt. Lett.* 41, 2069–2072 (2016).

[54] A. Annoni, et al., "Unscrambling light-automatically undoing strong mixing between modes," *Light Sci. Appl.* 6, e17110 (2017).

[55] Y. Yan, et al., "High-capacity millimetre-wave communications with orbital angular momentum multiplexing," *Nat. Commun.* 5, 4876 (2014).

[56] L. Gong, Q. Zhao, H. Zhang, et al., "Optical orbital-angular-momentum-multiplexed data transmission under high scattering," *Light Sci. Appl.* 8, 27 (2019).

[57] M. Cui, "Parallel wavefront optimization method for focusing light through random scattering media," *Opt. Lett.* 36, 870–872 (2011).

[58] S. Ramachandran, P. Kristensen, and M. F. Yan, "Generation and propagation of radially polarized beams in optical fibers," *Opt. Lett.* 34(16), 2525–2527 (2009).

[59] R. Beravat, G. K. L. Wong, M. H. Frosz, X. M. Xi, and P. S. J. Russell. "Twist-induced guidance in coreless photonic crystal fiber: A helical channel for light," *Sci. Adv.* 2, e1601421 (2016).

[60] P. S. J. Russell, R. Beravat, and G. K. L. Wong, "Helically twisted photonic crystal fibers," *Phil. Trans. R. Soc. A* 375, 20150440 (2017).

[61] X. M. Xi, G. K. L. Wong, M. H. Frosz, F. Babic, G. Ahmed, X. Jiang, T. G. Euser, and P. S. J. Russell, "Orbital-angular-momentum-preserving helical Bloch modes in twisted photonic crystal fiber," *Optica* 1, 165–169 (2014).

[62] C. N. Alexeyev, A. N. Alexeyev, B. P. Lapin, G. Milione, and M. A. Yavorsky, "Spin-orbit interaction-induced generation of optical vortices in multihelicoidal fibers," *Phys. Rev. A* 88, 063814 (2013).

[63] E. V. Barshak, C. N. Alexeyev, B. P. Lapin, and M. A. Yavorsky, "Twisted anisotropic fibers for robust orbital-angular-momentum-based information transmission," *Phys. Rev. A* 91, 033833 (2015).

[64] S. Longhi, "Bloch dynamics of light waves in helical optical waveguide arrays," *Phys. Rev. B* 76, 195119 (2007).

6 Tapered Chalcogenide Optical Fiber for Coherent Mid-IR Supercontinuum Generation

6.1 INTRODUCTION

In recent years, mid-IR supercontinuum light has captivated a lot of attention because of the manifestation of unique absorption bands of most of the molecules in this region [1]. Furthermore, mid-IR supercontinuum light is expected to have numerous prospective applications including flow cytometry, bio-photonic diagnostics, nonlinear spectroscopy, and infrared imaging [2–5]. In this direction, several optical fibers and waveguide structures have been reported for broadband mid-IR supercontinuum generation [6–18]. In most of the potential applications, we require intense, coherent, compact, and broadband mid-IR supercontinuum light sources [2, 19–21]. Supercontinuum light generated from a photonic crystal fiber was employed in a typical single-oscillator multiplex coherent anti-Stokes Raman scattering (CARS) step for the measurement of vibrational bands of cyclohexane sample [22]. Supercontinuum light generated from the integrated photonic chip can be used in mid-IR gas spectroscopy for the detection of gas such as acetylene (C_2H_2) [23]. For some high-spatial-resolution imaging applications, a spatially coherent supercontinuum light source is desirable. It has been demonstrated that optical fibers are a promising medium for the design and development of highly coherent mid-IR light sources with high brightness. The coherent mid-IR SCG has been reported using various optical fibers fabricated in different materials including silica, fluoride, tellurite, and chalcogenide glasses. In general, the coherence property of the supercontinuum spectrum depends on the input pump conditions, fiber geometrical parameters, and the mechanism involved in spectral broadening [24–26]. For example, the pump sources may be either femtosecond, picosecond, or nanosecond. The geometrical parameters of the optical fibers can be set in such a way that it exhibits either normal or anomalous dispersion at the operating wavelength. Another possibility is whether the optical fiber exhibits an all-normal dispersion profile or it offers both normal as well as anomalous dispersion profile in the interested region of bandwidth. The properties of the generated supercontinuum spectrum (*i.e.* bandwidth and coherence) would be different in different cases mentioned above. The different mechanisms

DOI: 10.1201/9781003502401-6

of generating supercontinuum spectrum are involved in different conditions of the pump and dispersion profile of the optical fibers.

In this chapter, emphasis will be given on the coherent SCG in all-normal dispersion (ANDi) engineered tapered optical fiber pumped with picosecond and femtosecond laser pulses. In the case of femtosecond laser pumping, the optical noise caused by the soliton fission and the modulation instability processes is entirely absent in an ANDi-engineered optical fibers [25]. In the optical fibers offering ANDi characteristic, the broadening mechanism of generating supercontinuum is mainly governed by the two nonlinear effects including self-phase modulation (SPM) and the optical wave breaking (OWB) [26].

A broadband supercontinuum spectrum spanning 960–2500 nm can be achieved using an ANDi chalcogenide (As_2S_3) microwire of only 3 mm length [27]. The microwire can be coated with PMMA and tapered to a diameter of 0.58 µm to achieve the ANDi regime. To generate supercontinuum spectrum, this was the first report on chalcogenide tapered microwire designed with an ANDi profile. The mechanism of the generation of supercontinuum is the OWB, a low-noise nonlinear process. Another experimental demonstration showed that a coherence mid-IR supercontinuum spectrum expanding from 2.2 µm to 3.3 µm can be generated using 2 cm long chalcogenide ($AsSe_2$) microstructured optical fiber pumped by femtosecond laser pulses at the wavelength of 2.7 µm [28]. In this structure of the microstructured optical fiber, four air holes around the core were replaced by the As_2S_5 glass rods. The refractive index of the As_2S_5 glass is lesser than the refractive index of $AsSe_2$ chalcogenide glass.

The numerical study on the tapered step-index chalcogenide fibers with various structural parameters was reported for the coherent mid-IR supercontinuum spectrum [29]. The characteristics of the pump laser considered in the numerical investigation were 500 femtosecond laser pulses of peak power of 1 kW at 4 µm. In this study, the bandwidth and the spectrally averaged coherence of the supercontinuum spectra at the output of tapered fiber with different length ratios (the length ratio is defined as the ratio of the transition region length near the laser input to the other transition region length near the output) under the same pumping conditions. Based on this numerical study, it is concluded that as the length ratio increases the bandwidth decreases from 4.84 µm to 4.11 µm while the spectrally averaged coherence increases from 0.53 to 0.9 and then jitters near the maximum. The length ratio within the range of 1–1.5 is desirable to preserve a balance between bandwidth and the coherence property of the generated spectrum.

Later, a tapered fluorotellurite microstructured optical fiber was reported for the demonstration of coherent supercontinuum light extending from 1.4 to 4 µm [30]. In this experiment, 4 cm long tapered fluorotellurite microstructured optical fiber and a 1980 nm femtosecond fiber laser were employed. The spectral broadening in the tapered fluorotellurite microstructured optical fiber is caused by the SPM, the Raman soliton, and red-shifted dispersive wave generation. The obtained results indicate that such types of fibers are promising nonlinear mediums for generating coherent broadband mid-infrared supercontinuum light.

Recently, a tellurite-based-chalcogenide tapered optical fiber (with $Ge_{20}As_{20}Se_{15}Te_{45}$ core and $Ge_{20}As_{20}Se_{20}Te_{40}$ cladding glasses) was fabricated via an isolated stacked extrusion technique [31]. The waist diameter and the length of the tapered optical fiber can be precisely controlled by a homemade tapering system. The tellurite-based chalcogenide glass tapered fiber is able to generate an ultra-broadband supercontinuum spectrum with high coherence characteristics. Such tellurite-based chalcogenide glass tapered fiber with a core diameter at the waist region of ≤14 μm offers ANDi characteristics in the spectral bandwidth of 1.7–14 μm. The coherent supercontinuum spectrum extending from 1.7 to 12.7 μm can be obtained using 7-cm long all-normal dispersion engineered tellurium-based-chalcogenide tapered optical fiber pumped at 5.5 μm. A robust chalcogenide glass nanotaper is able to generate a mid-IR supercontinuum spectrum extending over one octave of bandwidth (850 nm to 2.35 μm) by pumping of picosecond laser at 1.55 μm [32]. The 480 nm core diameter provides high nonlinearity, while a thermally compatible incorporated polymer jacket provides the nanotaper mechanically stable.

An efficient mid-IR supercontinuum spectrum can be generated in an arsenic tri-sulphide nano-spike waveguide embedded in silica when it is pumped with 65 fs laser pulses at 2 μm [33]. The schematic representation of an arsenic trisulphide nano-spike waveguide is illustrated in Figure 6.1(a) and longitudinal view of the nano-spike waveguide are shown in Figure 6.1(b). As illustrated in Figure 6.1(a), section A shows the nano-spike used for efficient coupling. The supercontinuum is generated in the constant-diameter part (section B). The tapered design of the 'nano-spike' transforms the incident light adiabatically into the fundamental mode of a 2-mm long uniform section with a diameter of 1 μm. This can enhance the launch efficiency. The nano-spike is fully encapsulated in a fused silica cladding, which protects it from the environment. Therefore, such type of nano-spikes provides a suitable means of launching light into sub-wavelength scale waveguides. Ultrashort (65 fs) laser pulses at a repetition rate of 100 MHz at 2 μm can be launched with an efficiency ~12% into the sub-wavelength nano-spike waveguide. The nonlinear mechanisms including soliton fission and the dispersive wave generation along the uniform section result in spectral broadening out to almost 4 μm for launched energies of only 18 pJ. The method of nano-spike offers a unique means of solving the

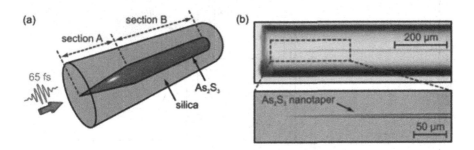

FIGURE 6.1   (a) The schematic representation of the nano-spike chalcogenide-silica step-index waveguide. (b) longitudinal image of the nano-spike (taken with an optical microscope) used in the experiments. The core diameter increases from 0 to 1 μm along the 300 μm long taper transition.

Adapted from [33].

long-standing problem of proficiently coupling light into the single-mode high-index contrast waveguides of sub-wavelength diameters.

The coherent mid-IR SCG in step-index tellurite, tapered tellurite, and chalcogenide double-clad optical fibers pumped with femtosecond laser system were also reported in the literature [34–36]. It has been found that chalcogenide glass fibers are much more suitable candidates for mid-IR supercontinuum applications. In the prior reported outcomes, the coherent supercontinuum spectra in chalcogenide fibers were achieved using the pumping at the longer wavelengths which are not conventionally accessible laser systems. Furthermore, the chalcogenide glasses offer large two-photon absorption with pumping at the longer wavelengths. However, the pumping at shorter wavelengths noticeably decreases the two-photon absorption phenomenon and sanctions operating in the normal dispersion region. The coherent mid-IR super-continuum spectrum using the chalcogenide fibers pumped at shorter wavelengths (< 2.7 μm) was not demonstrated earlier in the work demonstrated in Ref. [37]. This is because of that the achievable zero-dispersion wavelength of chalcogenide glass step-index fibers lies in a higher wavelength range. However, various new designs of the specialty optical fiber have been reported to tune the dispersion characteristic of the chalcogenide fibers to get the desired dispersion characteristics. The numerical report on chalcogenide-based M-type fiber structure is reported obtaining the ZDW of the core-confined higher-order mode (LP_{0n}) in the spectral range of 2 to 3 μm [38]. The refractive index profile of the designed M-type fiber with characteristic parameters is shown in Figure 6.2. M-type fiber structures have the excellent property that the higher-order (LP_{0n}) modes are core-confined and can be easily excited, while the fundamental LP_{01} and other modes are confined to a high-index ring surrounding the core, so they are not certainly excited. Such fibers are expected to have potential

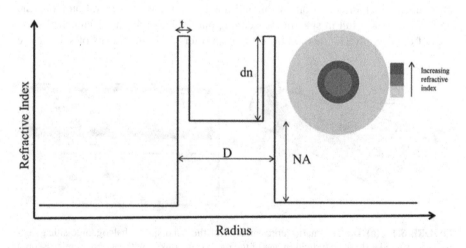

FIGURE 6.2   The refractive index profile of the M-type fiber with characteristic parameters t, d_n, D, and NA. The inset shows the end facet of an M-type fiber.

Adapted from [38].

applications in the future mid-IR supercontinuum sources pumped with established laser pump technology.

An annular-core PCF was reported for the novel guiding regime supporting wavelength-independent mono-annular guided modes (i.e. doughnut-shaped modes) [39]. In this design of PCF, only fundamental radial order modes are supported at all operating wavelengths. This unlocks the opportunity to attain the broadest and purest fiber-based supercontinuum vortex light supported by fiber eigenmodes of the fundamental radial order. This characteristic is of high interest for such applications that entail the stable and broadband guiding of mono-radial cylindrical vector beams and vortex beams carrying orbital angular momentum (OAM). The designed annular-core PCF structure illustrated in Figure 6.3(a) resembles that of a standard hexagonal lattice PCF having air holes with a diameter (d), pitch (Λ), and a number of rings of air holes (N). The main alteration though is that in the case of the annular-core PCF waveguiding occurs within a 'ring' of the six missing holes; while in the case of the standard PCF the optical modes are guided around a missing centre air hole. As shown in Figure 6.3(b), the annular-core PCF exhibits analogous waveguiding characteristics such as a fundamental HE_{11} mode showing an annular intensity profile along with other cylindrical vector modes TE_{0m}, HE_{2m}, TM_{0m} and other higher-order modes EH_{1m} and HE_{3m} with radial order m = 1. The intensity and phase distribution of $OAM_{\pm 11}$ modes supported in the annular-core PCF through the coherent

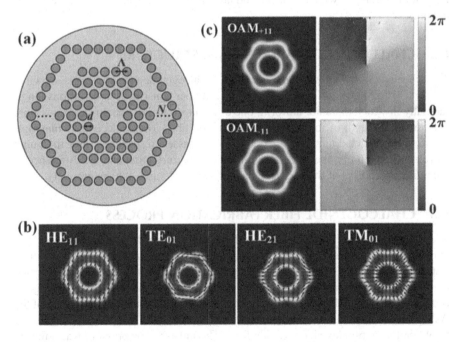

FIGURE 6.3 (a) The transverse cross-sectional view of annular-core PCF (b) The visualization of the intensity profiles of the first 4 guided vector modes. (c) Intensity and phase distributions of the $OAM_{\pm 11}$ modes supported by the annular-core PCF.

Adapted from [39].

superposition of hybrid modes: $OAM_{\pm 11} = HE_{21}(even) \mp i\ HE_{21}(odd)$ is illustrated in Figure 1(c). It is noted here that the other higher-order OAM modes can be supported by the fiber such as $OAM_{\pm 21} = HE_{31}(even) \mp i\ HE_{31}(odd)$ and $OAM_{\pm 21} = EH_{11}(even) \pm i\ EH_{11}(odd)$. Such annular-core PCF design has potential applications in optical sensing, space-division multiplexing, and super-resolution microscopy.

A tapered suspended-core fiber in a high-purity chalcogenide structure was designed and developed for the application of broadband mid-IR SCG using a standard fiber laser pump at 2 μm [40]. One of the advantages of the tapering of the fiber is that the length of the fiber for supercontinuum generation can be prominently reduced. It was revealed that a microstructured structure allows shifting a zero-dispersion wavelength to the range shorter than 2 μm in the fiber waist with a core diameter of about 1 μm. Using this chalcogenide tapered fiber, supercontinuum generation in the range of 1–10 μm can be obtained with 150-fs laser pulses of 100-pJ at 2 μm. Also, the experimental results illustrate that the wavelength conversion of ultrashort optical pulses at 1.57 μm from Er: fiber laser system in the chalcogenide tapered fibers can be achieved.

A chalcogenide tapered fiber with $AsSe_2$ as a core and As_2S_5 as a cladding glass has been fabricated for near-to-mid-IR supercontinuum generation [37]. Fabricated fiber offers an ANDi profile up to the wavelength of ~4 μm. The coherent mid-IR supercontinuum spectrum is demonstrated using the fabricated ANDi chalcogenide tapered fiber pumped with a femtosecond laser system at 2.0 to 2.6 μm (in the step of 0.2 μm).

6.2   CHALCOGENIDE TAPERED FIBER STRUCTURE

The schematic of the longitudinal view of the chalcogenide tapered fiber is shown in Figure 6.4. As shown in Figure 6.4, the tapered fiber consists of five sections longitudinally in which three are uniform, one down-tapered, and one up-tapered region. L is the total length of the tapered chalcogenide fiber. The chalcogenide materials for the core and cladding are $AsSe_2$ and As_2S_5 glasses, respectively. For the pump wavelength at 2.6 μm, the calculated numerical aperture (NA) and the difference in the refractive indices of the core and cladding are ~1.50 and ~0.5, respectively.

6.3   CHALCOGENIDE FIBER FABRICATION PROCESS

A chalcogenide fiber in a step-index refractive-index profile with $AsSe_2$ glass as a core and As_2S_5 glass as cladding was fabricated by the rod-in-tube method. As illustrated in Figure 6.5(a), firstly, $AsSe_2$ glass rod and As_2S_5 glass tube were fabricated by the casting and rotational casting method, respectively. After that the $AsSe_2$ glass rod was elongated and inserted into an As_2S_5 glass tube. Then the arrangement of the $AsSe_2$ rod and As_2S_5 tube was lengthened simultaneously using a fiber drawing tower to obtain the step-index chalcogenide fiber. The nitrogen gas pressure was adjusted as negative to avoid any interstitial hole forming between the core and cladding during the complete process of the fiber drawing unit. Finally, tapering of the fabricated step-index chalcogenide optical fiber was performed by employing a homemade fiber tapering system. The bench of the fiber tapering system was allowed to be inclined at an angle of 5 degrees from the horizontal position. We set the operating temperature

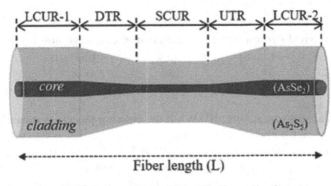

LCUR-1   : Large Core Uniform Region-1
DTR   : Down Tapered Region
SCUR   : Small Core Uniform Region
UTR   : Up Tapered Region
LCUR-2   : Large Core Uniform Region-2

FIGURE 6.4   The longitudinal view of the chalcogenide tapered optical fiber structure.

Adapted from [37].

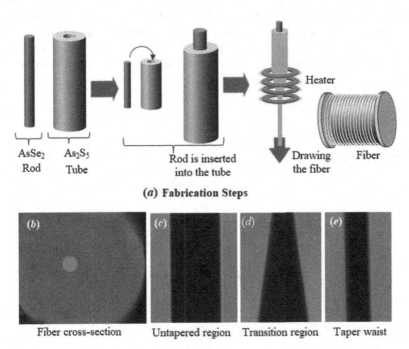

FIGURE 6.5   (*a*) The steps in the fabrication process of the chalcogenide step-index optical fiber using rod-in-tube method; (*b*) Cross-section of fabricate fiber; (*c*) Untapered region; (*d*) Transition region; (*e*) Taper waist.

Adapted from [37].

TABLE 6.1

The Geometrical Parameters of the Fabricated Chalcogenide Tapered Fiber

Geometrical	LCUR-1	SCUR	DTR / UTR	L
Parameters	0.7 cm	3 mm	4 mm	3.0 cm

of the circular filament at ~180 °C. When the tapering process is done the tapered fiber geometrical parameters were measured using a digital imaging system (Nikon: DS-5M-L1) for microscope (Nikon: ECLIPSE ME600L) and illustrated in Figure 6.5(b–e). The numerical values of the measured tapered geometrical parameters are provided in Table 6.1. The core diameter at the ends of the fiber was 15 μm and the minimum core diameter at the tapered waist was measured as 7 μm.

The chromatic dispersion characteristic of the tapered chalcogenide optical fiber with a core diameter varying from 7 μm to 15 μm is shown in Figure 6.6. It leads that the chalcogenide fiber offers ANDi profile up to the spectral range of 3.95 μm for the fibers with the core size varying from 7 μm to 15 μm. At the wavelength of 2.6 μm, the simulated chromatic dispersion values of the fiber with core diameters of 7 μm and 15 μm are -106.5 ps/nm.km and -125.4 ps/nm.km, respectively.

The spectral variations of the effective mode area of the fundamental mode propagating in the chalcogenide fiber with the core sizes of 7 μm and 15 μm are illustrated in Figure 6.7. The numerical values of effective mode area of the fundamental mode of the fiber with a core size of 7 μm and 15 μm are 20.85 μm² and 89.45 μm², respectively, at the wavelength of 2.6 μm. The wavelength-dependent nonlinear coefficient of the fundamental mode of the chalcogenide fiber is depicted in Figure 6.8.

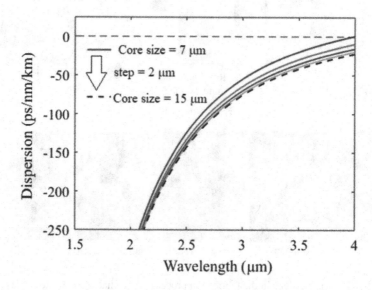

FIGURE 6.6 Dispersion profile of the fundamental propagating mode of the fiber with various core sizes varying from 7 μm to 15 μm with the step of 2 μm.

Adapted from [37].

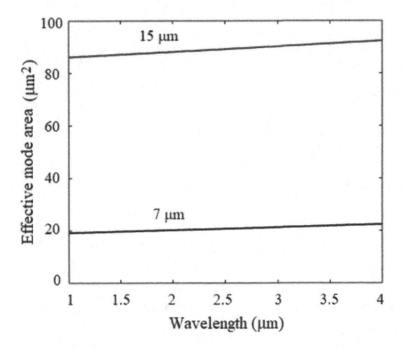

FIGURE 6.7 The spectral variations in the effective mode area of the fundamental mode of the fiber with core sizes of 7 μm and 15 μm.

Adapted from [37].

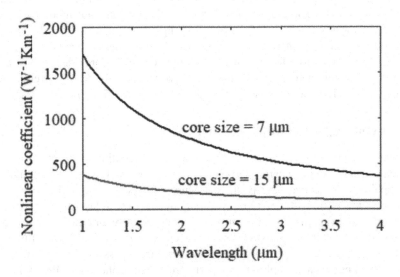

FIGURE 6.8 The spectral variations in the nonlinear coefficient of the fundamental mode of the fiber with core sizes of 7 μm and 15 μm.

Adapted from [37].

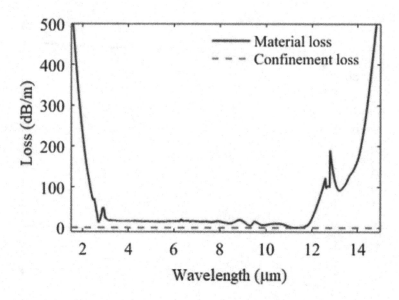

FIGURE 6.9   The material loss of AsSe$_2$ glass, and the confinement loss of the fiber.
Adapted from [37].

The numerical values of the nonlinear coefficient of the fiber with a core size of 7 μm
and 15 μm are 602.7 W^{-1} Km^{-1} and 141 W^{-1} Km^{-1} at 2.6 μm.

The material loss of the AsSe2-based chalcogenide glass and the confinement loss
of the fiber are illustrated in Figure 6.9. The absorption peak at ~2.9 μm is due to the
OH impurity in the glass, and another absorption peak at ~12.7 μm exists because of
Se-OH bonds [41]. The confinement losses of the fundamental mode have been
obtained at various core sizes of the fiber varying from 7 μm to 15 μm. It is a well-
known fact that the confinement loss increases by decreasing the core size. In Figure
6.9, the confinement loss for the fiber with 7 μm core size is provided. As shown in
Figure 6.9, the confinement loss of the fiber is very less in the whole spectral range
even at the smallest core diameter at the tapered region (i.e. 7 μm).

6.4   EXPERIMENTAL SETUP

The experimental setup for the measurement of supercontinuum spectra using the
fabricated chalcogenide tapered step-index fiber is shown in Figure 6.10. An ultrafast
Ti-sapphire mode-locked laser (Coherent Mira-900-F) delivering pulses at 800 nm
with a spectral pulse width of 12 nm was employed as a seed laser source. The seed
laser source delivers pulses to the pulse picker regenerative amplifier. The output of
the amplifier provides the pulses with a pulse energy of ~1 mJ at the repetition rate
of 1000 Hz. Then the amplified laser pulse permits to pass through the travelling
wave optical parametric amplifier of superfluorescence (Coherent TOPAS-C) which
generates a signal beam of 1.16–1.6 μm and an idler beam tunable from 1.6 to 2.6 μm
with the pulse width of 200 fs. A long-pass filter was employed to isolate the signal

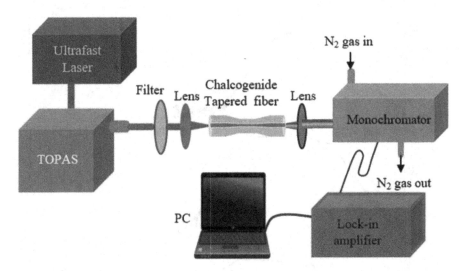

FIGURE 6.10 The illustration of the experimental setup used in the measurement of the supercontinuum spectrum from the chalcogenide tapered fiber.

Adapted from [37].

and idler beams. In our experiment, an idler beam at 2.0 to 2.6 μm wavelengths was allowed to couple into the 3 cm chalcogenide tapered fiber using an aspheric lens (THORLABS, C021TME-D, AR 1.8–3.0 μm) with a focal length of 11 mm. The transmission efficiency of the lens used in the experiment was approximately 60%. The estimated coupled peak power to the fiber was around 10.12 kW at 2.6 μm. The output of the chalcogenide tapered fiber was collected using a ZnSe lens with a focal length of 12 mm. The output spectra were measured using a monochromator (Bunkoukeiki CT-25) with 2 nm resolution. The transmission range of the ZnSe lens was 0.6–21.0 μm. The signal received from the monochromator was amplified using a lock-in-amplifier (NF LI5640) [35 – 37]. Finally, the supercontinuum spectrum was recorded using a computer-based spectrometer system.

6.5 RESULTS AND DISCUSSION

In the experiment, to characterize the fabricated fiber, a 3 cm long tapered chalcogenide fiber was used for the generation of the supercontinuum spectrum. The measured spectral broadening at the output end of the tapered chalcogenide fiber is shown in Figure 6.11 for various pump wavelengths of 2.0 μm, 2.2 μm, 2.4 μm, and 2.6 μm. Figure 6.8 shows the dependence of the measured supercontinuum spectra from ANDi tapered chalcogenide fiber pumped with 200 fs laser pulses of estimated coupled peak power of 11.88 kW, 11.60 kW, 11 kW, and 10.12 kW at pump wavelengths of 2.0 μm, 2.2 μm, 2.4 μm, and 2.6 μm, respectively. The measured supercontinuum spectral broadening spanned only from 1.5 to 2.6 μm for the pumping at 2.0 μm. When the pumping wavelength increased to 2.6 μm, the long-wavelength edge of the supercontinuum spectrum extended and the widest supercontinuum spectrum was

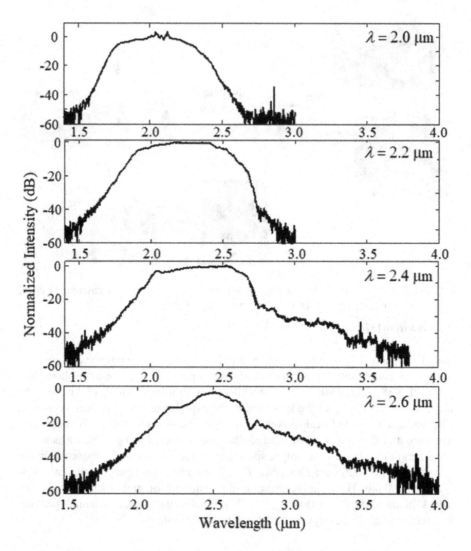

FIGURE 6.11   The variations in the broadening of supercontinuum spectra at the output of 3 cm long tapered chalcogenide fiber pumped with different pump wavelengths in the mid-IR region.

Adapted from [37].

generated. At the pump wavelength of 2.6 μm, the maximum spectral broadening extending from 1.6 μm to 3.7 μm has been measured. For the pumping in a normal dispersion regime, initially, SPM is the dominant nonlinear effect in the broadening mechanism of supercontinuum generation. Thereafter, OWB and higher-order dispersion are responsible for blue and redshifts of the spectrum.

The reported damage threshold of the As_2Se_3 glass in terms of laser fluence threshold (F_{th}) is 88.4 mJ/cm^2 for 150 fs laser pulses at 3 μm [42]. In the case of the experimental setup and laser pulse parameters, the maximum allowable peak power should be less than 3.52 MW to avoid fiber damage. In the actual experiment, the estimated coupled peak power is always less than the laser threshold of the chalcogenide glass. In support of the experimentally obtained result, the supercontinuum spectrum in tapered chalcogenide fiber with the same fiber geometrical parameters and pump conditions have been simulated numerically and discussed in the next section.

6.6 NUMERICAL METHOD AND ANALYSIS

To compare the measured supercontinuum spectrum, we performed numerical simulation by solving the following generalized nonlinear Schrodinger equation [43]

$$\frac{\partial \tilde{A}'}{\partial z} = i\bar{\gamma}(\omega)\exp\left(-\hat{L}(\omega)z\right)\mathcal{F}\left\{\bar{A}(z,T)\int_{-\infty}^{\infty} R(T')\left|\bar{A}(z,T-T')\right|^2 dT'\right\} \qquad (6.1)$$

The Eq. (6.1) was solved by employing the adaptive step-size method with the fourth-order-Runge-Kutta algorithm [43]. In the numerical simulations, the geometrical parameters of tapered chalcogenide fiber were taken as given in Table 6.1. The material and the confinement losses of the fiber have been included in all simulations.

The coherence characteristic of the generated supercontinuum spectrum is affected by the existence of the quantum noise of the pulse. We used one-photon-per-mode semi-classical theory to model the noise of the input pulse [44]. The complex degree of coherence was used to consider the deficit in the coherence characteristic of the supercontinuum spectrum due to the spectral phase instability at each wavelength. The relation for the complex degree of coherence is as follows [45]

$$\left|g_{12}^{(1)}(\lambda, t_1 - t_2)\right| = \left|\frac{\langle E_1^*(\lambda, t_1) E_2(\lambda, t_2)\rangle}{\sqrt{\langle\left|E_1(\lambda, t_1)\right|^2\rangle}\sqrt{\langle\left|E_2(\lambda, t_2)\right|^2\rangle}}\right| \qquad (6.2)$$

where E_1 and E_2 are the amplitudes of the electric field for two successive generated spectra. = 1 for completely coherent spectrum, and = 0 for entirely incoherent light.

The simulated spectral broadening of the supercontinuum spectrum from a 3.0 cm long tapered chalcogenide fiber, for the pump wavelengths varying from 2 μm to 2.6 μm, is illustrated in Figure 6.12(a–d). In the simulations, the peak power of the pump at various wavelengths has been considered the same as it was estimated in the experiment at a particular wavelength. The simulated and measured spectral broadening of the supercontinuum spectra is almost similar. In the ANDi fiber, the SPM and the OWB play a very important role in the spectral broadening of the supercontinuum spectrum. In the phenomena of SPM and OWB new components of the wavelengths with a phase related to the input, the pulse is created, and the noise-sensitive soliton dynamics are suppressed [46]. Therefore, the supercontinuum spectrum maintains the coherence characteristic in the ANDi fibers. As shown in Figure 6.12(e), the complex degree of coherence is almost unity (which corresponds to the perfect coherence) within the whole range of the generated supercontinuum spectrum. In the ANDi fiber pumped with ultrafast laser pulses, the nonlinear coupling between

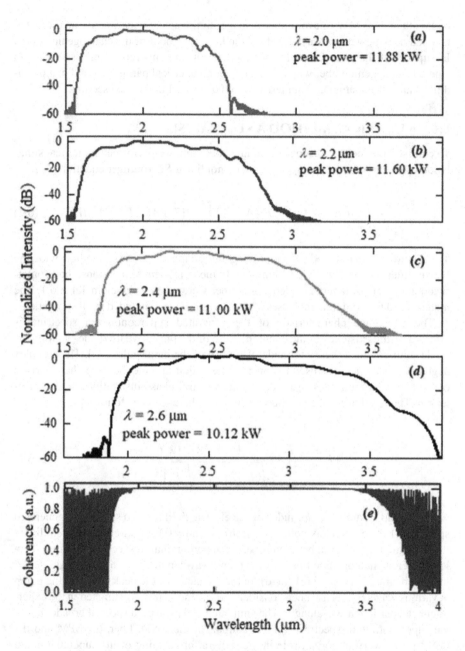

FIGURE 6.12   (a–d) Simulated supercontinuum spectra at the output of 3 cm long tapered chalcogenide fiber pumped at 2.0 to 2.6 μm laser pulse, (e) the coherence property of generated supercontinuum spectrum at 2.6 μm.

Adapted from [37].

nonlinear effects contributes to the suppression of incoherent dynamics and as a result, a highly coherent supercontinuum spectrum can be obtained [47].

The variations in the spectral broadening of the supercontinuum spectrum on laser pulse peak powers are illustrated in Figure 6.13. It is clear that when the peak power of the input pulse is 15 kW, the supercontinuum spectrum extends up to 4.5 micron at the higher wavelength edge. Therefore, in the experiment, the spectral broadening beyond 3.7 µm is limited by the input power coupled to the core of the fiber.

The coherent mid-IR supercontinuum generation in tapered soft-glass optical fibers is discussed. The coherent mid-IR supercontinuum generation using a tapered chalcogenide fiber pumped with femtosecond laser pulses at various wavelengths ranging from 2.0 µm to 2.6 µm has been extensively discussed experimentally as well as numerically. The experimentally measured results indicate that the coherent broadband supercontinuum spectrum spanning ~1.6 µm to 3.7 µm can be achieved using a very short (i.e. 3 cm) length of the tapered chalcogenide step-index optical fiber when it is pumped by the 200 fs laser pulses of peak power of 10.12 kW at 2.6 µm. Numerically simulated results support the experimentally measured results. Such highly coherent mid-IR supercontinuum light has potential applications in various fields including early cancer diagnostics, inspecting food quality, gas sensing, high-spatial-resolution imaging, and multiplex CARS microspectroscopy.

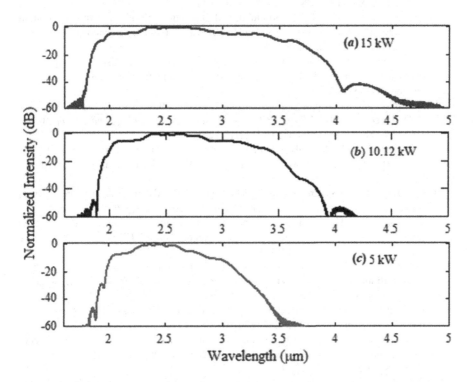

FIGURE 6.13 Variations in the supercontinuum spectra when the fiber is pumped by the laser pulses of the peak power of (a) 15 kW, (b) 10.12 kW, and (c) 5 kW at 2.6 µm.

Adapted from [37].

REFERENCES

[1] A. Schliesser, N. Picque, and T. W. Hansch, "Mid-infrared frequency combs," *Nat. Photon.* 6, 440–449 (2012).

[2] A. Labruyere, A. Tonello, V. Couderc, G. Huss, and P. Leproux, "Compact supercontinuum sources and their biomedical applications," *Opt. Fiber Technol.* 18, 375–378 (2012).

[3] R. Su, et al., "Perspectives of mid-infrared optical coherence tomography for inspection and micrometrology of industrial ceramics," *Opt. Express* 22, 15804–15819 (2014).

[4] F. C. Cruz, et al., "Midinfrared optical frequency combs based on difference frequency generation for molecular spectroscopy," *Opt. Express* 23, 26814–26824 (2015).

[5] M. Verdonck, et al., "Characterization of human breast cancer tissues by infrared imaging," *Analyst* 141, 606–619 (2016).

[6] S. S. Bobba and A. Agrawal, "Ultra-broad mid-IR supercontinuum generation in single, bi and tri layer graphene nano-plasmonic waveguides pumping at low input peak powers," *Sci. Rep.* 7, 10192 (2017).

[7] T. S. Saini, A. Kumar, and R. K. Sinha, "Broadband mid-infrared supercontinuum spectra spanning 2 – 15 µm using As_2Se_3 chalcogenide glass triangular-core graded-index photonic crystal fiber," *IEEE/OSA J. Lightwave Technol.* 33(18), 3914–3920 (2015).

[8] O. P. Kulkarni, et al., "Supercontinuum generation from ~1.9 – 4.5 µm in ZBLAN fiber with high average power generation beyond 3.8 µm using a thulium-doped fiber amplifier," *J. Opt. Soc. Am. B* 28(10) 2486–2498 (2011).

[9] I. Kubat, C. S. Agger, P. M. Moselund, and O. Bang, "Mid-infrared supercontinuum generation to 4.5 um in uniform and tapered ZBLAN step-index fibers by direct pumping at 1064 or 1550 nm," *J. Opt. Soc. Am.* 30(10), 2743–2757 (2013).

[10] T. S. Saini, A. Kumar, and R. K. Sinha, "Design and modeling of dispersion engineered rib waveguide for ultra broadband mid-infrared supercontinuum generation," *J. Modern Opt.* 64(2), 143–149 (2017).

[11] D. Jain, et al., "High power, ultra-broadnand supercontinuum source based on highly GeO_2 doped silica fiber," *Proc. SPIE, Fiber Lasers XIV: Technology and System*, 10083, 1008318 (2017). https://doi.org/10.1117/12.2251648

[12] J. Swiderski and M. Michalska, "High-power supercontinuum generation in a ZBLAN fiber with very efficient power distribution towards the mid-infrared," *Opt. Lett.* 39(4), 910–913 (2014).

[13] C. Agger, et al., "Supercontinuum generation in ZBLAN fibers-detailed comparison between measurement and simulation," *J. Opt. Soc. Am. B* 29(4), 635–645 (2012).

[14] T. S. Saini, U. K. Tiwari, and R. K. Sinha, "Design and analysis of dispersion engineered rib waveguides for on-chip mid-infrared supercontinuum," *IEEE/OSA J. Lightwave Technol.* 36(10), 1993–1999 (2018).

[15] L. Liu, G. Qin, Q. Tian, D. Zhao, and W. Qin, "Numerical investigation of mid-infrared supercontinuum generation up to 5 µm in single mode fluoride fiber," *Opt. Express* 19, 10041–10048 (2011).

[16] I. Kubat and O. Bang, "Multimode supercontinuum generation in chalcogenide glass fibers," *Opt. Express* 24, 2513–2526 (2016).

[17] C. R. Petersen, et al., "Mid-infrared supercontinuum covering the 1.4 – 13.3 µm molecular fingerprint region using ultra-high NA chalcogenide step-index fiber," *Nat. Photon.* 8, 830–834 (2014).

[18] T. S. Saini, et al., "Coherent mid-infrared supercontinuum generation using rib waveguide pumped with 200 fs laser pulses at 2.8 µm," *Appl. Opt.* 57(7), 1689–1693 (2018).

[19] S. Dupont, et al., "IR microscopy utilizing intense supercontinuum light source," *Opt. Express* 20(5), 4887–4892 (2012).

[20] S. L. Girard, M. Allard, M. Piche, and F. Babin, "Differential optical absorption spectroscopy lidar for mid-infrared gaseous measurements," *Appl. Opt.* 54(7), 1647–1656 (2015).

[21] M. N. Islam, et al., "Field tests for round-trip imaging at a 1.4 km distance with change detection and ranging using a short-wave infrared super-continuum laser," *Appl. Opt.* 55(7), 1584–1602 (2016).

[22] M. Falconieri, et al., "Characterization of supercontinuum generation in a photonic crystal fiber for uses in multiplex CARS microspectroscopy," *J. Raman Spectrosc. 50*, 1287–1295 (2019). https://doi.org/10.1002/jrs.5599

[23] D. Grassani, et al., "Mid-infrared gas spectroscopy using efficient fiber laser driven photonic chip-based supercontinuum," *Nat. Commun.* 10, 1553 (2019).

[24] K. L. Corwin, et al., "Fundamental amplitude noise limitations to supercontinuum spectra generated in a microstructured fiber," *Appl Phys B* 77, 269–277 (2003).

[25] M. Klimczak, et al., "Coherent supercntinuum generation up to 2.3 μm in all-solid soft-glass photonic crystal fibers with flat all-normal dispersion," Opt. Express. 22(15), 18824 – 18832 (2014).

[26] A. M. Heidt, "Pulse preserving flat-top supercontinuum generation in all-normal dispersion photonic crystal fibers," *J. Opt. Soc. Am. B* 27, 550–559 (2010).

[27] A. Al-Kadry, et al., "Broadband supercontinuum generation in all-normal dispersion chalcogenide microwires," *Opt. Lett.* 40(20), 4687–4690 (2015).

[28] L. Liu, et al., "Coherent mid-infrared supercontinuum generation in all-solid chalcogenide microstructured fibers with all-normal dispersion," *Opt. Lett.* 41(2), 392–395 (2016).

[29] L. Liu, K. Nagasaka, G. Qin, T. Suzuki, and Y. Ohishi, "Coherence property of mid-infrared supercontinuum generation in tapered chalcogenide fibers with different structures," *Appl. Phys. Lett.* 108, 011101 (2016).

[30] N. Li, et al., "Coherent supercontinuum generation from 1.4 to 4 μm in a tapered fluorotellurite microstructured fiber pumped by a 1980 nm femtosecond fiber laser," *Appl. Phys. Lett.* 110, 061102 (2017).

[31] N. Zhang, et al., "Ultrabroadband and coherent mid-infrared supercontinuum generation in Te-based chalcogenide tapered fiber with all-normal dispersion," *Opt. Express* 27(7), 10311–10319 (2019).

[32] S. Shabahang, "Octave-spanning infrared supercontinuum generation in robust chalcogenide nanotapers using picosecond pulses," *Opt. Lett.* 37(22), 4639–4641 (2012).

[33] N. Granzow, et al., "Mid-infrared supercontinuum generation in As_2S_3-silica 'nanospike' step-index waveguide," *Opt. Express* 21(9), 10969–10977 (2013).

[34] K. Nagasaka, et al., "Supercontinuum generation in chalcogenide double-clad fiber with near zero-flattened normal dispersion profile," *J. Opt.* 19, 095502 (2017).

[35] T. S. Saini, et al., "Coherent mid-infrared supercontinuum spectrum using a step-index tellurite fiber with all-normal dispersion," *Appl. Phys. Express* 11, 102501 (2018).

[36] T. S. Saini, et al., "Tapered tellurite step-index optical fiber for coherent near-to-mid-IR supercontinuum generation: Experiment and modeling," *Appl. Opt.* 58, 415–421 (2019).

[37] T. S. Saini, T. H. Tuan, T. Suzuki, and Y. Ohishi, "Coherent mid-IR supercontinuum generation using tapered chalcogenide step-index optical fiber: experiment and modelling," *Sci. Rep.* 10, 2236 (2020).

[38] D. Jain, C. Markos, T. M. Benson, A. B. Seddon, and O. Bang, "Exploiting dispersion of higher order-modes using M-type fiber for application in mid-infrared supercontinuum generation," *Sci. Rep.* 9, 8536 (2019). https://doi.org/10.1038/s41598-019-44951-4

[39] M. Sharma, P. Pradhan, and B. Ung, "Endlessly mono-radial annular core photonic crystal fiber for the broadband transmission and supercontinuum generation of vortex beams," *Sci. Rep.* 9, 2488 (2019). https://doi.org/10.1038/s41598-019-39527-1

[40] E. A. Anashkina, V. S. Shiryaev, M. Y. Koptev, B. S. Stepanov, and S. V. Muravyev, "Development of As-Se tapered suspended-core fibers for ultra-broadband mid-IR wavelength conversion," *J. Non-Crystalline Solids* 480, 43–50 (2018).

[41] T. Kohoutek, J. Orava, A. Greer, and H. Fudouzi, "Sub-micrometer soft lithography of a bulk chalcogenide glass," *Opt. Express*, 21(8), 9584–9591 (2013).

[42] C. You, et al., "Mid-infrared femtosecond laser induced damages in As_2S_3 and As_2Se_3 chalcogenide glasses," *Sci. Rep.* 7, 6497 (2017).

[43] J. Dudley and R. Taylor, *Supercontinuum generation in optical fibers*, Cambridge University Press (New York, 2010), 32–51.

[44] M. H. Frosz, "Validation of input-noise model for simulations of supercontinuum generation and rogue waves," *Opt. Express* 18, 14778–14787 (2010).

[45] J. M. Dudley, G. Genty, and S. Coen, "Supercontinuum generation in photonic crystal fiber," *Rev. Mod. Phys.* 78(4), 1135–1184 (2006).

[46] L. E. Hooper, P. J. Mosley, A. C. Muir, W. J. Wadsworth, and J. C. Knight, "Coherent supercontinuum generation in photonic crystal fiber with all-normal group velocity dispersion," *Opt. Express* 19(6), 4902–4907 (2011).

[47] A. M. Heidt, J. S. Feehan, J. H. V. Price, and T. Feurer, "Limits of coherent supercontinuum generation in normal dispersion fibers," *J. Opt. Soc. Am. B* 34(4), 764–775 (2017).

7 Applications of the Supercontinuum Generation

7.1 INTRODUCTION

The lasers are the most suitable light sources because they deliver an intense beam of coherent and collimated light. The narrowband spectral output from the laser source is greatly demanding. However, for some potential applications including medical imaging and testing of the optoelectronic devices for telecom networks, we required a broadband light source. The recent development of commercial supercontinuum sources based on optical fibers has gained much attention. Such light sources make use of various optical nonlinear effects in specifically designed optical fibers to generate light with a wide spectrum that can span the UV-to-visible-to-mid-IR. In this chapter, various applications of the broadband supercontinuum light generated in optical fibers are deliberated. The supercontinuum light sources are being used for various potential applications, including medical, biotechnology, military, and industry. In the medical field, the mid-IR supercontinuum light is applicable in coherence tomography, mid-IR spectroscopy, and bio-molecular sensing. The supercontinuum light sources are very suitable for detection of the cancer cells and their diagnostics. Ultrahigh-resolution optical coherence tomography using continuum generation in an air–silica photonic crystal fiber has already been demonstrated. It is also possible to show in-vivo multi-nonlinear optical imaging of a living cell using a supercontinuum light source generated from a photonic crystal fiber. For the defence sector, the mid-IR supercontinuum light offers potential applications in sensing and detecting explosive materials. Using the supercontinuum light sources, the white-light confocal microscope is conceivable for the spectrally resolved multidimensional imaging. It is possible to study photochemistry in the photonic crystal fiber-based nano-reactors using the supercontinuum light source.

7.2 SUPERCONTINUUM LIGHT FOR OPTICAL COHERENCE TOMOGRAPHY

Optical coherence tomography (OCT) is an imaging modality. Basically, OCT is used for micrometre-scale resolution imaging of the cross-sectional subsurface of the biological tissues. OCT is a non-invasive imaging test that employs broadband light sources and a fiber-optic interferometer to perform in-vivo high-resolution imaging of microstructure in biological tissues, including the retina [1]. It captures 3-D images of the optical scattering media. OCT provides ultra-sensitive and

DOI: 10.1201/9781003502401-7

high-resolution imaging in-vivo measurement of singly backscattered light as a function of depth. Using a Michelson interferometer, coherence-domain ranging in OCT can be possible. The longitudinal resolution, governed by the coherence length of the light source, is inversely proportional to the bandwidth of the light. In practice, the super luminescent diode (SLD) is being used for OCT imaging. The longitudinal resolution of OCT imaging using SLD-based light sources is 10–15 mm because of the limitation in optical bandwidth [2]. Using a Kerr-lens mode-locked Ti:sapphire laser can attain a longitudinal resolution OCT imaging less than 2 µm [3]. Recent developments in the specialty optical fiber technology allow an extremely broadband continuum of light sources using highly nonlinear photonic crystal fibers. Coherent supercontinuum light in an air–silica microstructure optical fiber can be utilized to demonstrate an ultrahigh-resolution OCT [4]. Using the developed broadband OCT system, it is possible to perform imaging with the bandwidth of 370 nm at an operating wavelength of 1.3 µm. Such OCT systems are capable of longitudinal resolutions of 2.5 µm in air and around 2 µm in biological tissue. With the developments in broadband light source technology, the longitudinal resolution has been enhanced significantly since OCT was first realized to image the eye's microstructure. However, with the advancement in the OCT resolution because of increment in the source bandwidths, one potential restriction that arises is the dispersion of the specimen medium. The low-dispersion light sources are advantageous for ultrahigh-resolution OCT imaging. The dependence of dispersion on depth is a limitation when scanning across a long distance, such as in the eye. Usually, OCT systems need adjustment of the dispersion stability between both arms of an interferometer [5]. However, this compensation works only for one plane, which averts high axial resolution over the complete scanning depth. Any unbalanced dispersion between two arms of the interferometer may destroy the axial resolution, and dynamic dispersion compensation is required for deep imaging depth to obtain ultrahigh-resolution OCT. The dispersion compensation based on post-processing of the OCT images with the known scanning-independent dispersion inconsistency has been demonstrated [6]. Nevertheless, such a dispersion compensation method requires substantial data processing. Another possibility is to tilt the grating in the frequency domain optical delay line (FD-ODL) to attain depth-dependent dispersion [7]. However, the limited bandpass of FD-ODL makes it problematic to be used in an ultra-high-resolution OCT. The impacts of the depth-dependent dispersion by the key component of the water, biological tissues, on the resolution of OCT have been investigated. It is conceivable to eradicate the influence of the depth-dependent dispersion in the tissue of water by selecting a light source with a centre wavelength near 1.0 µm. An ultra-broadband light source based on photonic crystal fiber with an operating wavelength of 1.0 mm can be used to perform ophthalmic imaging with a longitudinal resolution of 1.3 µm [8, 9].

The OCT system used in Ref. [9] was optimized to support 1.3 µm longitudinal resolution in biological tissues. The lateral resolution of the OCT system was 7 µm. The lateral resolution was determined by the achromatic objective lens. The anterior chamber of the rabbit eye was imaged in vitro. During the experiment, the central and limber regions of the eye were scanned. Figure 7.1(a) indicates the image of the corneal epithelium at the centre of a rabbit eye. The size of the image is 0.3 mm × 0.2 mm.

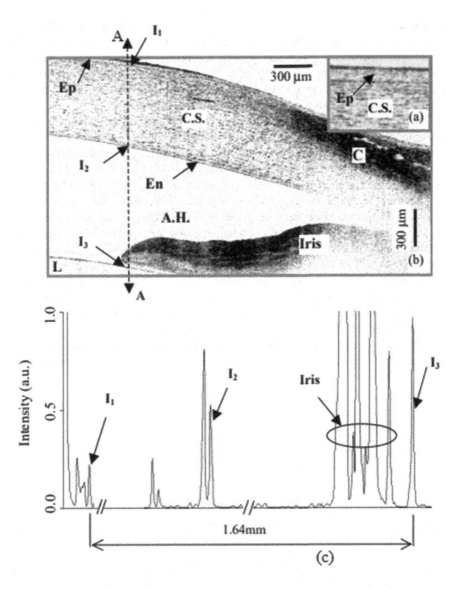

FIGURE 7.1 High-resolution ophthalmic imaging. (a) Cornea: image size 0.3 x 0.2 mm; Ep, epithelium; C.S., corneal stroma. (b) Anterior chamber of eye: image size 3.0 × 1.74 mm; En, endothelium; A.H., aqueous humour; C, conjunctiva; L, lens; I_1, interface between epithelium and stroma; I_2, interface between endothelium and A.H.; I_3, Interface between A.H. and lens. (c) Interference peaks at line A-A in (b).

Adapted from Ref. [9].

Figure 7.1(b) illustrate an image of the anterior chamber at the limbus of the eye to visualize the inner structures. The scanning rate of the moving stage was 2Hz. The time taken to get one image was recorded at about three minutes. The entire depth of the image is 1.74 mm, which was calibrated by the refractive index in the eye. The measured epithelium thickness at the limbus is about 39 μm, thinner than the thickness at the centre part. I_1 is the interface between the stroma and epithelium. I_2 represents the interface between the aqueous humour and endothelium. The iris, cornea, and conjunctiva can be clearly envisioned. I_3 is the interface between the lens and aqueous humour. As shown in the image, the width of the line at I_3 is uniform from left to right. Figure 7.1(c) illustrates the interference peaks at line A-A in (b).

The supercontinuum light enables us to execute unique multi-nonlinear optical imaging through various nonlinear optical processes. The SCG system can be used for OCT based on the tapered photonic crystal fiber [10]. A portable, compact, and powerful light source which can be deployed outside the laboratory environment, is suitable for the ultrahigh-resolution OCT imaging systems. The spectrally flat supercontinuum light can be produced in a tapered photonic crystal fiber pumped with a compact Ti:sapphire laser that releases relatively low energy (~2 nJ) femtosecond pulses at a wavelength of 809 nm. The advantage of tapering a PCF is that the dispersion profile can be controlled in the taper waist to obtain a spectrally flat continuum generated by self-phase modulation. The broadband and flat supercontinuum light can be generated with a bandwidth of up to 177 nm at 809 nm centre wavelength, at an output power of 108 mW (78% of input power). Using such flat and broadband supercontinuum light, ultrahigh-resolution OCT imaging is possible on selected material samples [10].

As shown in Figure 7.2, according to the full width at half maxima (FWHM) interference envelope of the signal the maximum accessible free-space resolution of

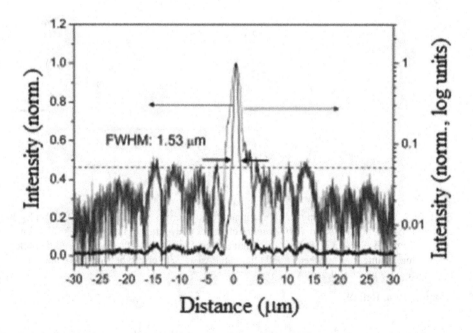

FIGURE 7.2   The envelope of the interference signal of the widest supercontinuum (black curve: linear scale, blue curve: log. scale) with the side-lobe level marked (dashed line).

Adapted from Ref. [10].

1.5 μm can be demonstrated [10]. The sidelobes observed in the autocorrelation function, caused by residual modulations in the spectrum, were equal to or less than 5% of the signal maximum. The longitudinal resolution of less than 1.5 μm can be obtained by increasing the output power of the laser. The convenience of this supercontinuum light source for the ultrahigh-resolution imaging has been confirmed by performing measurements on selected material samples in combination with en-face scanning OCT system [11]. The ultrahigh-resolution cross-sectional view and en-face OCT images of various polyolefin foam samples with average pore sizes of 500 μm and 80 μm, respectively, are shown in Figures 7.3(a–d). The demodulated interferogram from a single depth-or A-scan with the mirror as a sample is illustrated in Figure 7.3(e). In Figures 7.3(g, h), a 150 μm thick shielding coating with surrounded ceramic particles on wood has been imaged and compared to a light microscope image taken from a polished cross-section (Figure 7.3(f)).

The images of the fine particles in the coating, as well as small structures within the cell walls of the foam samples, are clearly resolved and demonstrate the utility of the supercontinuum light source for material investigation with ultrahigh-resolution OCT systems. Disparate in vivo OCT measurements that advantage from a broadband light source with a centre wavelength near 1 μm [9], for the study of nonliving substances the high power and high resolution of the current system, will be of advantage.

7.3   SCG FOR NONLINEAR OPTICAL IMAGING

The nonlinear optical imaging comprises a variety of optical phenomena, including second harmonic generation (SHG), two-photon excited fluorescence (TPEF), and coherent anti-Stokes Raman scattering (CARS). The principles of SHG, TPEF and CARS, are illustrated in Figure 7.4(a–c) through the energy level scheme of the processes. As shown in Figure 7.4(a), in case of the SHG, two photons with the same frequency (ω_1) interact with a nonlinear material, are "combined", and produce another photon with twice the frequency ($2\omega_1$) of the initial photons (equivalently, twice the energy and half the wavelength), that conserves the coherence of the excitation. It is a special case of sum-frequency generation (two photons), and more generally of harmonic generation.

Two-photon excitation fluorescence is the process in which a fluorophore (a molecule that fluoresces) is excited by the simultaneous absorption of two photons of an identical frequency (ω_1). This process normally involves photons in the ultraviolet or blue/green spectral region. Nevertheless, a similar excitation process can be created by the simultaneous absorption of two less energetic photons (typically in the infrared spectral range) under adequately intense laser illumination. This nonlinear phenomenon can take place if the sum of the energies of the two photons is greater than the energy gap between the ground states of the molecule and excited states. Subsequently, this process depends on the simultaneous absorption of two infrared photons, the possibility of two-photon absorption by a fluorescent molecule is a quadratic function of the excitation radiance. Under adequately intense excitation, three-photon and higher photon excitation is also conceivable and deep UV microscopy based on these processes has been developed. CARS is a third-order nonlinear optical effect concerning three laser beams: a pump laser beam of frequency ω_{pump}, a

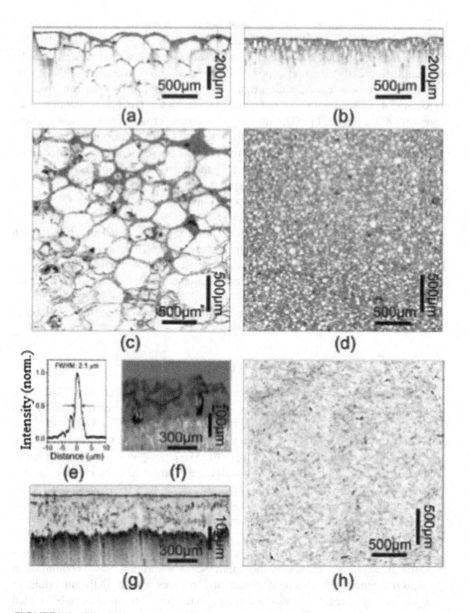

FIGURE 7.3 Ultrahigh-resolution OCT images of selected material samples: cross-sectional images of (a, b) polyolefin foams with different pore sizes, (g) a 150 μm thick protective coating on wood. (c, d, h) Corresponding en-face OCT images of the foam specimens and the coating, (f) microscope image of a polished cross-section of the protective coating. (e) Demodulated interferogram (from a single depth- or A-scan) with a mirror as a sample.

Adapted from Ref. [10].

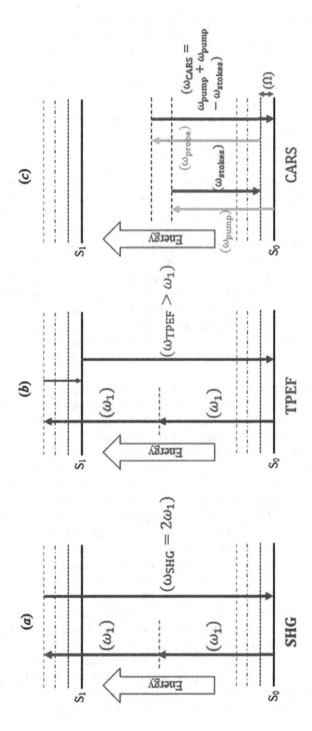

FIGURE 7.4 An energy diagram illustrating three best-known nonlinear optical effects used in biomedical imaging: (a) second harmonic generation (SHG); (b) two-photon excited fluorescence (TPEF); and coherent anti-Stokes Raman scattering (CARS).

Stokes beam of frequency ω_{stokes} and a probe beam at frequency ω_{probe}. All these beams interact with the sample and produce a coherent optical signal at the anti-Stokes frequency ($\omega_{pump}+\omega_{probe}-\omega_{stokes}$). The latter is resonantly enhanced when the frequency difference between the pump and the Stokes beams ($\omega_{pump}-\omega_{stokes}$) coincides with the frequency of a Raman resonance, which is the basis of the technique's intrinsic vibrational contrast mechanism. The CARS microscopy permits vibrational imaging with high speed, high sensitivity, and three-dimensional spatial resolution.

The CARS microscopy is a technique for creating three-dimensional images of living cells. In CARS microscopy two laser sources (*i.e.* pump and probe) are required, where the difference in their frequencies matches the resonance of the sample molecule to be studied. It allows us to image living cells without using toxic dyes.

In order to explain the intra-cellular structure and its dynamics *in vivo* with three-dimensional sectioning competency, Raman microspectroscopy is one of the most prevailing and non-destructive techniques [12–17]. Based on the time- and space-resolved molecular-specific data acquired by confocal Raman microspectroscopy, the mitochondrial metabolic activity and an initial death process of a living yeast cell have been investigated. Multiple vibrational resonances have been noticed through the CARS process. Owing to an ultra-broadband spectral profile of the supercontinuum light, multiple vibrational resonances can be detected. An ultra-broadband multiplex CARS microspectroscopy has been demonstrated using a coherent supercontinuum in the near-infrared region generated from a photonic crystal fiber [18]. Owing to the ultra-broadband Stokes radiation attained from the supercontinuum light, multiple vibrational modes can be excited simultaneously in the wave-number range of more than 2500 cm^{-1}.

The supercontinuum light sources permit us to execute unique multi-nonlinear optical imaging through various nonlinear optical processes. A supercontinuum light produced using a femtosecond Ti:Sapphire oscillator has been employed to acquire both the vibrational and TPEF images of the living cell concurrently at different wavelengths. Owing to an ultra-broadband spectral profile of the supercontinuum, multiple vibrational resonances have been detected through the CARS process. In addition to the multiplex CARS process, manifold electronic states can be excited because of the broadband electronic two-photon excitation using the supercontinuum light, giving rise to a TPEF signal. It is possible to visualize the organelles such as the septum, mitochondria, and nucleus through the CARS and the TPEF processes by employing a living yeast cell with the nucleus labelled by the green fluorescent protein (GFP) [19]. Therefore, the supercontinuum light enables us to perform unique multi-nonlinear optical imaging through two different nonlinear optical processes. The input pump laser pulses should be spectrally filtered using a narrow bandpass filter in order to achieve CARS spectrum with a high-frequency resolution. In Ref. [19], the bandwidth of the pump laser was around 20 cm^{-1}. The energy of the input pulse of the pump and Stokes lasers were 200 and 170 pJ, respectively. Both the laser pulses were superimposed collinearly using an 800-nm Notch filter, and then tightly focused onto the sample with a microscope objective (x40). The forward-propagating CARS signal was collected with another microscope objective (x40) in an opposed configuration. The characteristic spectral profiles of the CARS signal of a living yeast cell (red) and surrounding water (blue) obtained in 100 ms exposure time is shown in Figure 7.5(a). The CARS spectra of the living yeast cell (red) and water background (blue) are found at (x, y) = (5.65 μm, -3.05 μm) and (6.71 μm, 1.22 μm), which are marked as black and white crosses in Figure 7.5(b), respectively. As clearly shown in this figure, a strong signal is detected inside of the living yeast cell at the Raman shift of 2840 cm^{-1}. This band initiates from CH_2 stretching vibrational modes, which illustrates a slightly dispersive line shape due to an interference effect with the non-resonant background. In another study of supercontinuum-based multiplex CARS microspectroscopy, only a vibrational resonant signal was extracted using a differentiation method [20]. It is possible to extract only a vibrationally resonant CARS image from a typical spectral profile in the C–H stretching vibrational region. The vibrational contrast can be considerably improved in contrast to CARS imaging at a fixed Raman shift. Since, the mitochondrion is an organelle containing a high concentration of phospholipids, the signal at the Raman shift of 2840 cm^{-1} is found especially in mitochondria. The CARS image of the living cells at the Raman shift of 2840 cm^{-1} is shown in Figure 7.5(b). It indicates several yeast cells at various stages of the cell division cycle. Exclusively, a septum is envisioned in the yeast cell around the centre of Figure 7.5(b). The septum is composed of carbohydrates such as sac-charide, which is also rich in CH_2 bonds. The CARS signal of the yeast cell decreases at the edge because of the imperfect focusing due to the refractive index mismatch between the surrounding water and the yeast cell. Since the CARS microscopy offers the three-dimensional sectioning competency, it enables us to achieve not only a lateral but also an axial slice of the living yeast cell.

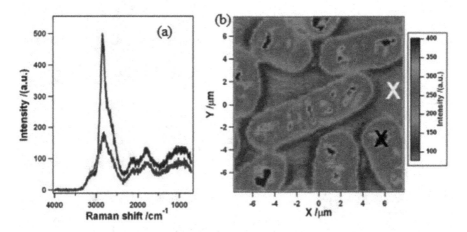

FIGURE 7.5 (a) The spectral profiles of the CARS signal of a living yeast cell (red) and surrounding water (blue) obtained in 100 ms exposure time; (b) CARS image of living yeast cell at a Raman shift of 2840 cm^{-1}. The CARS signal at the positions of black and white crosses correspond to the red and blue curves in Figure 7.5(a), respectively.

Adapted from Ref. [19].

The lateral and axial CARS images of a yeast cell are illustrated in Figures 7.6 (a) and 7.6(b), resembles the vertical slice of a yeast cell at the position of $y = 0$. It is clear that the CARS signal is feebler at the top portion rather than the bottom portion. It is due because of the deficient focusing of the two laser beams due to the spatially heterogeneous refractive index inside of the cell. The axial slices of the same yeast cell at each depth position are illustrated in Figure 7.6(c). From this figure, the three-dimensional spreading of the mitochondria is obviously observed with the high three-dimensional spatial resolution. Because of the broadband feature of the super-continuum light, the spectral characteristic of the multiplex CARS signal can be interpreted in detail. Along with the ultra-broadband multiplex CARS detection, the proficient two-photon excitation can also be executed in resonance with the two-photon indorsed electronic state. The Raman signal is caused by a nucleus and it is weak to be perceived, dual imaging of the CARS and the TPEF signals offers suitable information on the dynamical behaviour of the living system with high speed. The

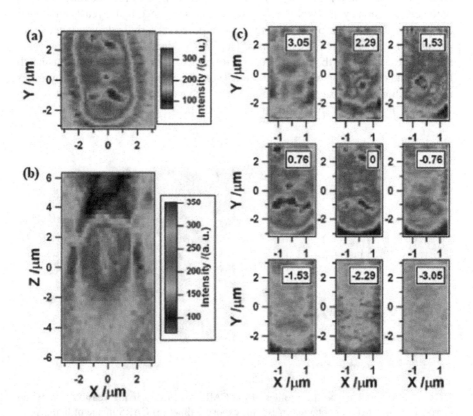

FIGURE 7.6   Lateral (a) and axial (b) CARS images of a yeast cell at a Raman shift of 2840 cm^{-1}; (c) Axial slices of the yeast cell at different depth positions, which are indicated at the top in a micrometre scale.

Adapted from Ref. [19].

supercontinuum light allows us to implement unique multi-nonlinear optical imaging through numerous nonlinear optical processes.

Nonlinear optical imaging is a developing technology with plenty of potential in pharmaceutical analysis [21]. The collective potential of these phenomena for pharmaceutical imaging includes chemical and solid-state specificity, high optical spatial and temporal resolution, non-destructive and non-contact analysis, no requirement for labels, and compatibility with imaging in aqueous and biological environments. The CARS microscopy can be utilized for imaging based on the C–H stretch vibrational region (2800–3200 cm^{-1}). For understanding the drug delivery mechanisms as well as designing tailor-made release profiles visualization of the 3-dimensional distribution of the drug molecules and sequential alterations during the release process is critical. This type of study utilized CARS imaging to inspect paclitaxel distribution in several polymer films with a lateral resolution of 0.3 mm and depth resolution of 0.9 mm [22]. Raman bands in the CH stretch vibration and fingerprint regions have been used to discriminate paclitaxel from the polymers. The detection sensitivity can be obtained to be 29 mM by imaging paclitaxel molecules dissolved in N, N-dimethylformamide solution. Ejection of paclitaxel from a polymer matrix can be monitored at an acquisition speed of 1 frame/s. In this way, CARS microscopy can be applied efficiently for in situ imaging of native drug molecules in the delivery system.

The time-dependent variations in the 3-dimensional distribution of paclitaxel in the polymer film have been envisioned by means of CARS microscopy [23]. CARS images revealed that the paclitaxel was distributed uniformly throughout the polyfilm styrene-b-isobutylene-b-styrene (SIBS) matrix. The alterations in the paclitaxel distribution during discharge were monitored using depth intensity profiles. This suggests that the quantitative CARS intensity of paclitaxel diminished upon exposure of the paclitaxel-loaded film to a release medium. These results show that paclitaxel was dissolved and depleted from the SIBS film during in vitro drug elution. This experiment supports the use of CARS microscopy as an effective non-destructive technique for chemical imaging of paclitaxel elution dynamics in polymer films. The CARS microscopy was also utilized to envisage the release of a model drug (theophylline) from a lipid (tripalmitin) based tablet during dissolution [24]. The effects of dissolution and transformation of the drug can be imaged in real-time. It was suggested that the manufacturing process causes substantial differences in the release process: tablets prepared from powder show the formation of theophylline monohydrate on the surface which prevents a controlled drug release, whereas solid lipid extrudates did not show the formation of monohydrate. This imagining technique can aid future tablet design. Spectrofluorometric imaging microscopy has been established in a confocal microscope by employing the supercontinuum light as an excitation source and a custom-built prism spectrometer for detection [25]. Such a microscope system delivers confocal imaging with spectrally resolved fluorescence excitation and detection from 450 to 700 nm. The supercontinuum light source delivers a broadband spectrum and can be coupled with an acoustic-optic tunable filter to offer continuously tunable fluorescence excitation with a 1-nm bandwidth. It is possible to select eight different excitation wavelengths simultaneously. The prism spectrometer offers spectrally resolved detection with sensitivity analogous to the standard confocal system. Such a new microscope system empowers optimum access to a multitude of fluorophores and delivers fluorescence excitation and emission spectra for each location in a 3D confocal image.

Real-time observation of the variations of chlorophyll distributions and cellular structures in the leaves during plant progression delivers significant information for understanding the physiological statuses of the plants. The SHG imaging and two-photon-excited autofluorescence imaging of leaves can be used for intensive care to study the nature of intrinsic fluorophores dissemination and cellular structures of the leaves by employing the near-infrared region of the light, which has minimal light absorption by endogenous molecules and thus upsurges tissue penetration. A highly polarized supercontinuum light generated from a birefringent nonlinear PCF with two zero-dispersion wavelengths can successfully excite two-photon autofluorescence and the SHG signals at the same time monitoring intrinsic fluorophore distributions and non-centrosymmetric structures of leaves [26]. Recently, the advances in UV supercontinuum generation, near-IR microscopy, exciting new capacities for the use of hollow-core PCFs, on-chip supercontinuum generation, and technologies to improve supercontinuum stability for certain applications have been reviewed [27].

7.4 SCG FOR SENSING APPLICATIONS

The recognition of the gas-phase molecular species at low concentration levels is a striking research field, offering a broad variety of environmental, industrial, and biological applications. Amongst a number of detection techniques, mid-IR laser absorption spectroscopy delivers an exceptional opening for sensitive and selective detection by targeting the typical spectral patterns in the so-called mid-IR fingerprint region. The supercontinuum light sources are well-matched for the gas sensing, as they deliver broadband spectral coverage in the near-IR and mid-IR regions, where numerous gaseous species and biomolecules of interest for bio/chemical processes or atmospheric sensing applications have vibrational overtone transitions. As illustrated in Figure 7.7, the simplest operation of an absorption measurement (based on the supercontinuum light source) is a direct absorption measurement using a spectrometer for detection. The supercontinuum light can be generated using nonlinear single-mode fiber pumped with an intense narrow-linewidth pulse laser. In this arrangement the high repetition rate light source can be regarded as a quasi-continuous wave source. The spectrometer employed for the detection would ideally feature both the high spectral resolution (to allow explicit identification of the spectral lines) and broadband spectrum (to permit whole spectral bands of one or more species to be recorded).

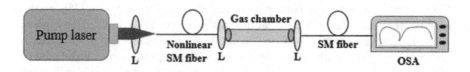

FIGURE 7.7 The simplest experimental setup for supercontinuum light-based absorption measurement (L: lens, SM: single mode, OSA: optical spectrum analyzer).

The study of one of the rotational-vibrational overtone bands of H_2O, extending from 1350 nm to 1500 nm, can serve as an example of the broadband capability of supercontinuum light-based sensors [28]. The onset of high-brightness and compact supercontinuum radiation sources using solid-core PCFs is opening to make an influence across the field of applied spectroscopy research. The applications of the supercontinuum light sources to build innovative instrumentations for the wavelength flexible hyperspectral confocal imaging, and chemical sensing were demonstrated [29].

The latest progressions of mid-IR supercontinuum light sources have opened up new opportunities in laser-based trace gas sensing. Whereas the supercontinuum light sources intrinsically support wide spectral coverage, the detection of broadband absorption signals with high speed and low cost is traditionally limited by the unavailability of the mid-IR detector arrays. This constraint can be sidestepped by up-converting the mid-IR signal into the near-IR region, where the inexpensive silicon-based detector arrays can be employed to measure broadband absorption. Furthermore, by combining a mid-IR supercontinuum light with a mid-IR-to-near-IR up converter and an astigmatic multi-pass cell, fast detection of ethane with sub-ppmv sensitivity can be accomplished at room temperature [30]. The basic experimental arrangement for the trace gas sensor is shown in Figure 7.8. The PPLN/CPLN crystal is placed inside the laser cavity demarcated by the dichroic mirrors (DM1 and DM3). In Figure 7.8, the green, blue, and red lines indicate the intra-cavity pump, near-IR output, and the mid-IR input beams, respectively. A mass-flow controller and the pressure controller are connected to the inlet and outlet of the multi-pass cell, respectively, in order to sustain a steady flow condition. The characterization of the spectral coverage and the resolution of the developed gas sensor has been carried out by targeting a number of (broadly absorbing) gas species. First of all, the mixture of gas of 50 ± 1.2 ppmv nitrous oxide (N_2O in N_2) and the same amount of ethylene (C_2H_4 in N_2) at a flow rate of 5 L/h was sent to the multi-pass cell. The constant pressure was stabilized at 900 mbar. Both the PPLN and CPLN crystals of the

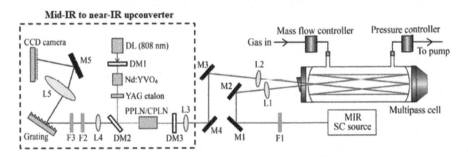

FIGURE 7.8   The schematic illustration of the trace gas sensor (M1-M5: mirrors; L1 and L2: 50 cm focal lenses; L3: 35 mm focal lens; L4: lens system of an effective focal length of 20 cm; L5: lens system of an effective focal length of 10 cm; DM1-DM3: dichroic mirrors; DL: diode laser; F1: 2.4 μm long-pass filter; F2: 1000 nm short-pass filter; F3: 750 nm long-pass filter).

Adapted from Ref. [30].

up-converter were utilized individually, and the associated absorbance spectra, each averaged for one minute, are shown in Figures 7.9(a & b), respectively. As illustrated in Figure 7.9(a & b), the absorption spectra obtained by experiment and the numerical simulations carried out based on HITRAN [31] and PNNL [32] databases were found in good agreement. For the broadband trace gas sensing, both the crystals (PPLN and CPLN) show brilliant perception. The up-conversion limitation at the lower frequency side is due to the cut-off of the SC power. The up-conversion restraint at the lower frequency side is due to the cut-off of the supercontinuum power. Towards the higher frequency side, the up-conversion restraint is because of numerous factors including the un-optimized reflectivity of the mirror of the multi-pass cell and inadequate supercontinuum power.

The enactment of the up-conversion-based trace gas sensor by focusing on a single species, i.e. ethane (C_2H_6) can be evaluated at low concentration levels using PPLN crystal. Figure 7.10 (a), illustrations an overlap of two measured absorbance spectra of ethane (10 ± 0.3 ppmv, 900 mbar), prepared by diluting gases from dedicated gas cylinders by means of controlled gas flow. Each spectrum was averaged for 20 ms. The projection of the absorbance values of the second measurement with respect to the first measurement is depicted in Figure 7.10(b). As demonstrated in Figure 7.10(b), the red line represents the linear fit with a slope value close to one which signifies good reproducibility.

The quantification of the multiple gas species with overlying absorption is more puzzling compared to single-species sensing. To explore this issue, the mixture of three gases including acetaldehyde, ethane, and ethylene (all, 100±2 ppmv, Linde Gas Benelux, Netherlands) has been examined. A target concentration of 33.33 ± 1.30 ppmv was applied to each of these species by means of the controlled mass flow. The evaluation was performed by spectrally decomposing the measured absorbance of the gas mixture into influences from the distinct species.

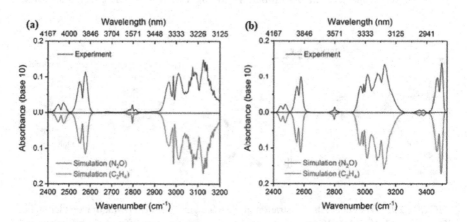

FIGURE 7.9 The absorbance spectra of the gas mixture of 50 ± 1.2 ppmv N_2O and 50 ± 1.2 ppmv C_2H_4 at 900 mbar realized by employing the PPLN crystal (a) and the CPLN crystal (b). The simulated spectra are inverted as the red (N_2O) and pink (C_2H_4) curves.

Adapted from Ref. [30].

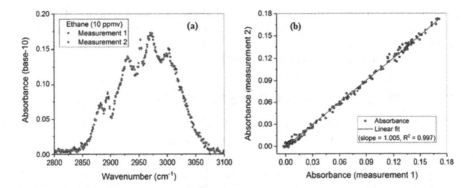

FIGURE 7.10   (a) An overlap of two absorbance spectra of 10 ppmv ethane at 900 mbar. Each measured for 20 ms. (b) Projection of the second ethane absorbance measurement with respect to the first measurement. The linear fit is included as the red line.

Adapted from Ref. [30].

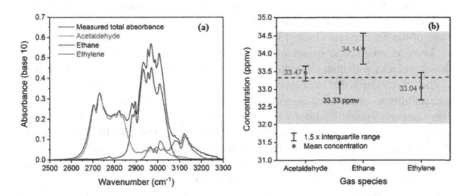

FIGURE 7.11   (a) The measured spectrum of absorbance (black curve) of the mixture of the gas averaged for 1 second. The primary curves represent the contributions from acetaldehyde (red). Ethane (blue) and ethylene (green) obtained by simulation based on the PNNL database. (b) Calculated concentration values by using the non-negative least-square curve fitting method. The associate interquartile ranges are included based on the statistics of 60 measurements. The targeted concentration of 33.33±1.30 ppmv is highlighted as the dashed blue line, comprising the dilution uncertainty represented by the grey area.

Adapted from Ref. [30].

The typically measured spectrum of absorbance (blue curve) of the gas mixture averaged for 1 second is depicted in Figure 7.11(a). The spectra of individual gases including acetaldehyde, ethylene, and ethane are simulated and represented by red, green, and blue curves, respectively. The calculated concentration values by using the non-negative least-square curve fitting method are shown in Figure 7.11(b). The associate interquartile ranges are included based on the statistics of 60 measurements. The targeted concentration of 33.33±1.30 ppmv is highlighted as the dashed blue line, comprising the dilution uncertainty represented by the grey area. Therefore, the least-square global fitting method can be provided for multi-species detection, presenting and encouraging potential for various applications such as biomedical research and environmental monitoring [30].

7.5 SUPERCONTINUUM LIGHT SOURCES FOR METROLOGY

The advances in the multi-watt broadband supercontinuum light sources promise polarized radiant flux with expended spectral coverage from visible to mid-IR. The supercontinuum light sources are very exciting sources for their potential applications in fundamental optical metrological [33–35]. Phase-locked white-light continuum pulses can be used to obtain a broadband frequency comb for entire frequency measurements from UV to IR. The communal phase coherence within the 'optical comb lines makes it conceivable to relate the microwave frequency or radiofrequency to an optical one in a single direct step. It has become forthright to determine the absolute frequencies of all of the comb lines. This capability has transformed optical frequency metrology and synthesis. The capacity to tally the optical oscillations of $>10^{15}$ cycles/second enables high-precision optical spectroscopy and has led to the building of an all-optical atomic clock that is expected ultimately to outstrip today's state-of-the-art caesium clocks [36].

There is an interest in achieving more precise measurements of frequency to develop clocks with instability down to ~10^{-17} at 1 second. It is feasible to have state-of-the-art standards that are based on the ions, atoms, and molecules to exhibit excellent stability to obtain ultra-high reproducibility and accurateness for the clocks. The 'femtosecond optical frequency comb generation is the new concept in clock technology. The optical pulse obtained from a stabilized mode-docked laser is better for the frequency comb generation. The femtosecond laser with the supercontinuum can have the proper mode spectrum to generate a broadband optical pulse with the spectrum extending from visible to near-infrared. These optical pulses are repeated at steady rates (100 MHz to 1 GHz, based on design), leading to an 'optical comb' of the frequencies with excellent phase coherence and stability. The optical combs contain some millions of steady coherent optical frequencies. The communal phase coherence within the optical comb lines makes it conceivable to relate the microwave frequency or radiofrequency to an optical one in a single direct step [37].

To understand the spectrum of an infinite train of broadband pulses one can imagine that the periodicity in the time domain will manifest itself as a periodicity in the frequency domain. However, the comb of the frequencies when extended to near-zero frequency will not match up with the repetition rate and the frequency zero. While, in the femtosecond laser, the cavity modes are well-defined by the phase velocity and the envelope repletion rate is fixed by the group velocity. Usually, the phase and group velocities differ from each other. Consequently, the details of the optical oscillations under each successive femtosecond pulse differ. Essentially, a phase slip between successive output pulses appears as an extra optical phase. Therefore, the general formula of a comb-line frequency is given by [38]:

$$f_n = f_{ceo} + nf_{rep} \qquad (7.1)$$

Here, f_n represents the optical frequency of comb-line n, and for 100 MHz repetition rate, n=5×10⁶. The career-envelope offset frequency is f_{ceo} and the phase slip per pulse (δ) is given as:

$$\delta = 2\pi \left(\frac{f_{ceo}}{f_{rep}} \right) \tag{7.2}$$

The optical heterodyning region of the supercontinuum yields additional stability for the measurement of frequency with a fractional frequency noise of the order of 6×10⁻¹⁶ in 1 second of averaging over the bandwidth of 300 THz. Further, the small changes in the gravity with the frequency accuracy of 10⁻¹⁸ can easily be noticed.

The evaluation of the potential for the supercontinuum light sources coupled to monochromators has been carried out for three simple radiometric metrology applications including measurements of filter transmittance, determination of irradiance responsivity, and the measurements of bidirectional reflectance distribution function (BRDF) [39]. It has been found that the spectral power density and the stability of the supercontinuum light source determine the range of promising metrological applications.

7.6   PCF-BASED SCG FOR MULTISPECTRAL PHOTOACOUSTIC MICROSCOPY

The high-energy SCG is suitable for multispectral photoacoustic microscopy. The high-power SCG can be achieved using a carefully designed large-core tapered PCF. The large-core input at the input side provides an improved power handling capability while the small-core at the tapered end provides the desired spectral range of the supercontinuum spectrum. Figure 7.12 shows the schematic of the longitudinal profile of a tapered PCF and its microscopic images of the cross-section of the end facets for taper 10.7–5 μm. The total longitudinal length of PCF was ~15 m. The length of the untampered straight input section is ~2 m and the length of the tapered section is 1 m for taper 10.7–5 μm and 0.5 m for 10.3–5 μm [40].

The experimental setup for the measurement of the supercontinuum spectrum is shown in Figure 7.13. An Elforlight laser (Model FQS-400-1-Y-1064) delivering pulses with a duration of 4 ns, a repetition rate of 1 kHz, and pulse energies of up to 500 μJ is used as a pump source operating at 1064-nm wavelength. A polarizing beam splitter (PBS) and a half-wave plate (HWP) permit for continuous adjustment of the laser power. An aspheric lens (AL) is employed to couple the pump laser into a PCF. In order to maximize the coupling efficiency, the focal length of AL is optimized for the PCF.

Finally, the supercontinuum spectra can be recorded using an optical spectrum analyzer. All the measurements can be plotted in terms of energy density in nJ/nm.

Experimental results showed that the tapered PCF allows high-power supercontinuum output pulse energy without shifting the blue edge of the spectrum from 475 nm. The experimentally measured supercontinuum spectrum has a pulse energy density above 15 nJ/nm over a high spectral bandwidth extending from 500 nm to 1600 nm. Such high-power broadband supercontinuum light sources are highly desirable for functional multispectral photoacoustic microscopy.

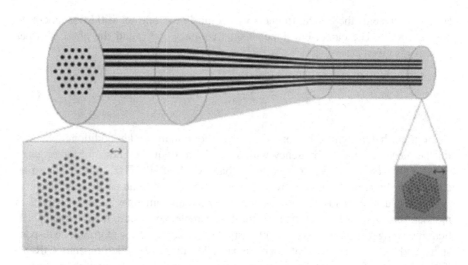

FIGURE 7.12 The schematic of the longitudinal profile of tapered fibers and its microscopic images of the cross-section of the untampered and tapered end facets.

Adapted from Ref. [40].

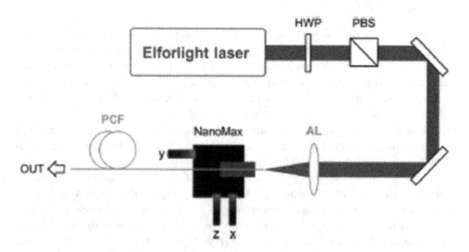

FIGURE 7.13 The experimental setup of the SCG using tapered PCF.

Adapted from Ref. [40].

7.7 FUSED TAPERED PCF-BASED SENSOR

A fused tapered PCF interferometer can be made based on the discontinuous cooling method. In the tapering process, stop heating and wait for cooling at different taper lengths. Repeat the heating process and cooling process, until the taper goes to the estimated length. Compared with the regular fused tapered method, the fringe contrast of the transmission spectra of this sensor can be obtained up to 15.06 dB.

The experimental setup for the fused tapered PCF-based sensor is illustrated in Figure 7.14. The transmission spectra in different concentrations of the glycerol solution are shown in Figure 7.15. The experimental results illustrate that as the external refractive index increases, the transmission spectra of the sensor shift towards the longer wavelengths. In the measurement of the glycerol solution, the refractive index sensitivity of the sensor reaches 797.674 nm/RIU, while, the obtained temperature sensitivity is only 0.00125 nm/ °C [41]. The results showed that the tapered PCF-based sensor is temperature insensitive and can overcome the cross-sensitivity problem of the simultaneous measurement of the refractive index and the temperature.

A compact PCF Mach–Zehnder modal interferometer was reported based on no adiabatically tapered special silica fibers coated with an ultra-thin BSA antigen or an 8 nm palladium film to detect the interaction between the BSA antigen and an anti-BSA antibody with a record detection limit of 125 pg/ml of the antibody concentration or for fast detection of low hydrogen concentrations, between 1.2 and 5.6 vol% [42]. Figure 7.16 illustrates the image of the tapered PCF. In the figure, L_0 represents the length of the taper waist and ρ_w indicates the waist diameter.

The interaction of the adiabatically tapered PCF fundamental mode with a thin film absorbing coating, deposited on the surface of the taper waist, on the transmission of the tapered PCF allows developing of the photonic biochemical sensors. It has been found that a prominent sensory effect occurs in the case of the resonant coupling between a fundamental mode and the cladding modes localized between PCF air channels and the absorbing coating. The temperature-independent refractometer can be fabricated by sandwiching a tapered PCF between two standard single-mode fibers. I have noticed that tapering the PCF greatly enhances the sensitivity of the refractometer [43]. The graphene-coated tapered PCF Mach–Zehnder interferometer can also be developed for sensing hydrogen sulfide gas [44]. The air

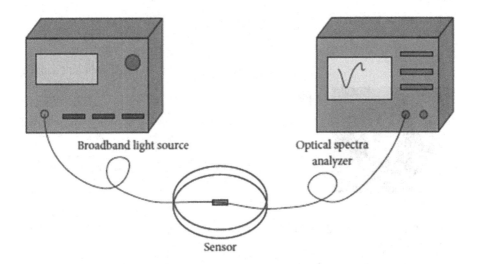

FIGURE 7.14 The experimental setup of the fused tapered PCF-based sensor.

Adapted from Ref. [41].

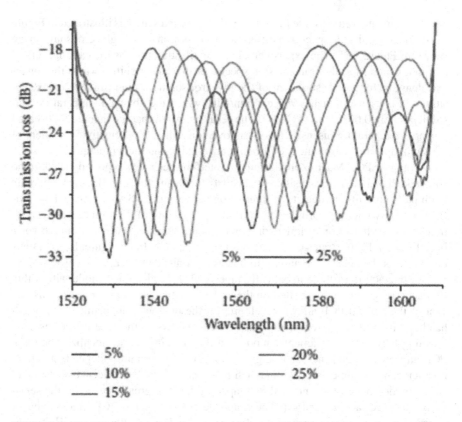

FIGURE 7.15 The transmission spectra of the tapered PCF-based sensor for the glycerol solution.

Adapted from Ref. [41].

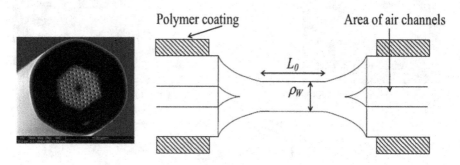

FIGURE 7.16 A micrograph of the PCF cross-section (left) and the illustration of the tapered PCF (right).

Adapted from Ref. [42].

holes of the PCF in the splicing regions are fully collapsed so it is favourable to the mode coupling. The tapered PCF Mach–Zehnder interferometer can be coated with a layer of graphene by using a dip-coating and sintering process. Such sensors have the advantages of simple structure, easy manufacture, low cost, and high sensitivity, which can be used in indoor gas sensing fields such as malls, laboratories, and factories. Also, a liquid crystal can be coated on tapered PCF for the development of an interferometer [45]. The properties of the liquid crystal materials (low and high indices) with the tapered PCF interferometers provide extra potential to make the multi-species sensing systems.

REFERENCES

[1] D. Huang, E. Swanson, C. P. Lin, J. S. Schuman, W. G. Stinson, W. Chang, M. R. Hee, T. Flotte, K. Gregory, C. A. Puliafito, and J. G. Fujimoto, "Optical coherence tomography," *Science* 254, 1178–1181 (1991).

[2] R. C. Youngquist, S. Carr, and D. E. N. Davies, "Optical coherence-domain reflectometry: A new optical evaluation technique," *Opt. Lett.* 12, 158–160 (1987).

[3] W. Drexler, U. Morgner, F. X. Kartner, C. Pitris, S. A. Boppart, X. D. Li, E. P. Ippen, and J. G. Fujimoto, "In vivo ultrahigh-resolution optical coherence tomography," *Opt. Lett.* 24, 1221–1223 (1999).

[4] I. Hartl, X. D. Li, C. Chudoba, R. K. Ghanta, T. H. Ko, J. G. Fujimoto, J. K. Ranka, and R. S. Windeler, "Ultrahigh-resolution optical coherence tomography using continuum generation in an air-silica microstructure optical fiber," *Opt. Lett.* 26(9), 608–610 (2001).

[5] C. K. Hitzenberger, A. Baumgartner, W. Drexler, and A. F. Fercher, "Dispersion effects in partial coherence interferometry: Implications for intraocular ranging," *J. Bio. Opt.* 4, 144–151 (1999).

[6] J. F. de Boer, C. E. Saxer, and J. S. Nelson, "Stable carrier generation and phase-resolved digital data processing in optical coherence tomography," *Appl. Opt.* 40, 5787–5790 (2001).

[7] E. D. J. Smith, A. V. Zvyagin, and D. D. Sampson, "Real-time dispersion compensation in scanning interferometry," *Opt. Lett.* 27, 1998–2000 (2002).

[8] Y. Wang, Y. Zhao, J. S. Nelson, Z. Chen, and R. S. Windeler, "Ultrahigh-resolution OCT using broadband generation from a photonic crystal fiber," *Opt. Lett.* 28, 182–184 (2003)

[9] Y. Wang, J. S. Nelson, Z. Chen, B. J. Reiser, R. S. Chuck, R. S. Windeler, "Optimal wavelength for ultrahigh-resolution optical coherence tomography," *Opt. Express* 11(12), 1411 (2003).

[10] G. Humbert, W. J. Wadsworth, S. Leon-Saval, J. C. Knight, T. Birks, P. S. J. Russell, M. Lederer, D. Kopf, K. Wiesauer, E. Breuer, and D. Stifter, "Supercontinuum generation system for optical coherence tomography based on tapered photonic crystal fiber," *Opt. Express* 14(4), 1596–1603 (2006).

[11] K. Wiesauer, M. Pircher, E. Götzinger, S. Bauer, R. Engelke, G. Ahrens, G. Grützner, C. K. Hitzenberger, and D. Stifter, "En-face scanning optical coherence tomography with ultra-high resolution for material investigation," *Opt. Express* 13, 1015–1024 (2005).

[12] G. J. Puppels, F. F. M. De Mul, C. Otto, J. Greve, M. Robert-Nicoud, D. J. Arndt-Jovin, and T. M. Jovin, "Studying single living cells and chromosomes by confocal Raman microspectroscopy," *Nature* 347, 301–303 (1990).

[13] Y. Takai, T. Masuko, and H. Takeuchi, "Lipid structure of cytotoxic granules in living human killer T lymphocytes studied by Raman microspectroscopy," *Biochim. Biophys. Acta* 1335, 199–208 (1997).

[14] C. Otto, N. M. Sijtsema, and J. Greve, "Confocal Raman microspectroscopy of the activation of single neutrophilic granulocytes," *Eur. Biophys. J.* 27, 582–589 (1998).

[15] Y.-S. Huang, T. Karashima, M. Yamamoto, and H. Hamaguchi, "Molecular-level pursuit of yeast mitosis by time- and space-resolved Raman spectroscopy," *J. Raman Spectrosc.* 34, 1–3 (2003).

[16] Y.-S. Huang, T. Karashima, M. Yamamoto, and H. Hamaguchi, "Molecular-level investigation of the structure, transformation, and bioactivity of single living fission yeast cells by time- and space-resolved Raman Spectroscopy," *Biochemistry* 44, 10009–10019 (2005).

[17] Y. Naito, A. Toh-e, and H.-O. Hamaguchi, "In vivo time-resolved Raman imaging of a spontaneous death process of a single budding yeast cell," *J. Raman Spectrosc.* 36, 837–839 (2005).

[18] H. Kano and H. Hamaguchi, "Ultrabroadband (>2500 cm^{-1}) multiplex coherent anti-stokes raman scattering microspectroscopy using a supercontinuum generated from a photonic crystal fiber," *Appl. Phys. Lett.* 86, 121113–121115 (2005).

[19] H. Kano and H. O. Hamaguchi, "In-vivo multi-nonlinear optical imaging of a living cell using a supercontinuum light source generated from a photonic crystal fiber," *Opt. Express* 14(7), 2798–2804 (2006).

[20] H. Kano and H. Hamaguchi, "Vibrationally resonant imaging of a single living cell by supercontinuum-based multiplex coherent anti-Stokes Raman scattering microspectroscopy," *Opt. Express* 13, 1322–1327 (2005).

[21] https://www.americanpharmaceuticalreview.com/Featured-Articles/148760-Nonlinear-Optical-Imaging-Introduction-and-Pharmaceutical-Applications/

[22] E. N. Kang, et al., In situ visualization of paclitaxel distribution and release by coherent antistokes Raman scattering microscopy. *Anal. Chem.* 78(23), 8036–8043 (2006).

[23] E. Kang, H. Wang, K. Kwon, Y. H. Song, K. Kamath, K. M. Miller, J. Barry, J. X. Cheng, and K. Park. "Application of coherent anti-stokes Raman scattering microscopy to image the changes in a paclitaxel–poly(styrene-b-isobutylene-b-styrene) matrix pre- and post-drug elution," *J. Biomed. Mat. Res.* A 87A(4), 913–920 (2008).

[24] M. Jurna, M. Windbergs, C. J. Strachan, L. Hartsuiker, C. Otto, P. Kleinebudde, J. L. Herek, and H. L. Offerhaus. "Coherent anti-Stokes Raman scattering microscopy to monitor drug dissolution in different oral pharmaceutical tablets," *J. Innov. Opt. Health Sci.* 2(1), 37–43 (2009).

[25] J. H. Frank, A. D. Elder, J. Swartling, A. R. Venkitaraman, A. D. Jeyasekharan, and C. F. Kaminski, "A white light confocal microscope for spectrally resolved multidimensional imaging," *J. Microsc.* 227(3), 203–215 (2007).

[26] W. Tao, H. Bao, and M. Gu, "Enhanced two-channel nonlinear imaging by a highly polarized supercontinuum light source generated from a nonlinear photonic crystal fiber with two zero-dispersion wavelengths," *J. Biomed. Opt.* 16(5), 056010 (2011).

[27] C. Poudel and C. F. Kaminski, "Supercontinuum radiation in fluorescence microscopy and biomedical imaging applications," *J. Opt. Soc. Am.* B 36(2), A139–A153 (2019).

[28] R. S. Watt, C. F. Kaminski, and J. Hult, "Generation of supercontinuum radiation in conventional single-mode fiber and its application to broadband absorption spectroscopy," *Appl. Phys.* B 90, 47 (2008).

[29] C. F. Kaminski, R. S. Watt, A. D. Elder, J. H. Frank, and J. Hult, "Supercontinuum radiation for applications in chemical sensing and microscopy," *Appl. Phys.* B 92(3), 367–378 (2008).

[30] K. E. Jahromi, Q. Pan, L. Hogstedt, S. M. M. Friis, A. Khodabakhsh, P. M. Moselund, and F. J. M. Harren, "Mid-infrared supercontinuum-based upconversion detection for trace gas sensing," *Opt. Express* 27(17), 24469–24480 (2019).

[31] L. S. Rothman, I. E. Gordon, Y. Babikov, A. Barbe, D. C. Benner, P. F. Bernath, M. Birk, L. Bizzocchi, V. Boudon, L. R. Brown, A. Campargue, K. Chance, E. A. Cohen, L. H. Coudert, V. M. Devi, B. J. Drouin, A. Fayt, J. M. Flaud, R. R. Gamache, J. J. Harrison, J. M. Hartmann, C. Hill, J. T. Hodges, D. Jacquemart, A. Jolly, J. Lamouroux, R. J. Le Roy, G. Li, D. A. Long, O. M. Lyulin, C. J. Mackie, S. T. Massie, S. Mikhailenko, H. S. P. Muller, O. V. Naumenko, A. V. Nikitin, J. Orphal, V. Perevalov, A. Perrin, E. R. Polovtseva, C. Richard, M. A. H. Smith, E. Starikova, K. Sung, S. Tashkun, J. Tennyson, G. C. Toon, V. G. Tyuterev, and G. Wagner, "The HITRAN2012 molecular spectroscopic database," *J. Quant. Spectrosc. Radiat. Transf.* 130, 4–50 (2013).

[32] T. J. Johnson, L. T. M. Profeta, R. L. Sams, D. W. T. Griffith, and R. L. Yokelson, "An infrared spectral database for detection of gases emitted by biomass burning," *Vib. Spectrosc.* 53(1), 97–102 (2010).

[33] T. Udem, R. Holzwarth, and T. W. Hansch, "Optical frequency metrology," *Nature* 416(6877), 233–237 (2002).

[34] N. Savage, "Supercontinuum sources," *Nat. Photon.* 3, 114–115 (2009). https://doi.org/10.1038/nphoton.2008.286

[35] R. Hontinfinde, S. Coulibaly, P. Megret, M. Taki and M. Wuilpart. "Metrology of supercontinuum generation along highly nonlinear fibers using photon-counting optical time domain reflectometry," *Proceeding: Conference on Lasers and Electro-Optics Pacific Rim (CLEO-PR)*, IEEE, Singapore, 31 July–4 August 2017. https://doi.org/10.1109/CLEOPR.2017.8118680

[36] M. Bellini and T. W. Hansch, "Phase-locked white-light continuum pulses: Toward a universal optical frequency-comb synthesizer," *Opt. Lett.* 25(14), 1049 (2000).

[37] S. A. Diddams, D. J. Jones, J. Ye, S. T. Cundiff, J. L. Hall, J. K. Ranka, R. S. Windeler, R. Holzwarth, T. Udem, and T. W. Hansch "Direct link between microwave and optical frequencies with a 300 THz femtosecond laser comb," *Phys. Rev. Lett.* 84, 5102–5105 (2000).

[38] J. L. Hall and J. Ye, "Optical frequency standards and measurement," *IEEE Trans. Instrum. Meas.* 52(2), 227 (2003).

[39] J. T. Woodward, A. W. Smith, C. A. Jenkins, C. Lin, S. W. Brown, and K. R. Lykke, "Supercontinuum sources for metrology," *Metrologia* 46, S277 (2009).

[40] M. M. Bondu, C. D. Brooks, C. Jakobsen, K. Oakes, P. M. Moselund, L. Leick, O. Bang, and A. Podoleanu, "High energy supercontinuum source using tapered photonic crystal fibers for multispectral photoacoustic microscopy," *J. Biomed. Opt.* 21(6), 061005 (2016).

[41] G. Fu, X. Fu, P. Guo, Y. Ji, and W. Bi, "Research on fused tapered photonic crystal fiber sensor based on the method of intermittent cooling," *J. Sens.* 2016, 7353067 (2016).

[42] V. P. Minkovich and A. B. Sotsky, "Tapered photonic crystal fibers coated with ultra-thin films for highly sensitive bio-chemical sensing," *J. Eur. Opt. Soc. Rapid Pub.* 15(7) (2019). https://doi.org/10.1186/s41476-019-0103-6

[43] K. Ni, C. C. Chan, X. Dong, C. L. Poh, and T. Li, "Temperature-independent refractometer based on a tapered photonic crystal fiber interferometer," *Opt. Commun.* 291, 238–241 (2013).

[44] X. Feng, W. Feng, C. Tao, D. Deng, X. Qin, and R. Chen, "Hydrogen sulfide gas sensor based on graphene-coated tapered photonic crystal fiber interferometer," *Sens. Actuat B Chem.* 247, 540–545 (2017).

[45] G. Rajan, S. Mathews, G. Farrell, and Y. Semenova, "A liquid crystal coated tapered photonic crystal fiber interferometer," *J. Opt.* 13, 015403 (2010).

8 Future Perspectives of the Photonic Crystal Fibers

8.1 INTRODUCTION

There would be growing demand for additional effective cleavers, low-loss splicers, multi-port couplers, intra-fiber devices, and mode-area transformers, etc. Consequently, through PCF-based technology, we are approaching superior technology, huge job opportunities, and an improved world. As yet unexplored is the use of twisted PCF in nonlinear optics and fiber lasers, where the combination of circular and orbital angular momentum (OAM) birefringence with control of group velocity dispersion may offer opportunities for new kinds of mode-locked soliton lasers, wavelength conversion devices, and powerful supercontinuum sources. The ability of twisted fibers to provide OAM and circular birefringence suggests that yet more possibilities may emerge from this unique and unexpected guidance mechanism. The merging of nanotechnology on the PCF is exploring lab-on-fiber technology. Several other future perspectives of the PCF are also elucidated in this chapter.

8.2 FUTURE PERSPECTIVES

The favourable demonstrations of the proof-of-principle applications of the supercontinuum sources for the basic metrological applications indicate numerous future directions of the photonic research to determine the possibility of upgrading the radiometric calibration facilities to take in the new types of sources. Some of the reported potentials include: (i) Estimate the uncertainty in irradiance responsivity calibration of colorimeters and photometers; (ii) Calculate the uncertainty budget for bidirectional reflectance distribution function measurements; (iii) Evaluate supercontinuum-monochromator sources for use with pyroelectric detectors for simplified scale extension to 2.5 μm; (iv) Estimate the prospective for the use in spectral radiance responsivity calibrations; and (v) Extend Zong's stray light correction algorithm [1] to the monochromators and estimate the dynamic range limit for out-of-band irradiance responsivity calibrations.

The calibration service facilities may be improved based on the final uncertainties attainable in the radiometric calibrations using the multi-channel detectors and the supercontinuum light sources. Nevertheless, some improvements are desirable before the supercontinuum sources can be completely exploited. Some of the suggested improvements include: (i) Polarization-preserving fiber output; (ii) Prolonged spectral output into the UV region; (iii) Response for the active power stabilization to reduce the noise in the output beam; and (iv) Unravelling visible and near-infrared

                                    DOI: 10.1201/9781003502401-8

spectral regions using a dichroic beam splitter. With some exclusion most of the above-listed anticipated improvements are presently accessible or will be obtainable soon in the commercial systems.

Commercially available white light supercontinuum sources are rapidly becoming an indispensable tool in laboratories globally and in commercial microscopes. The research, development, and use of the high-power industrial PCF-based laser are awe-inspiring day by day. The benefits of the PCF-based supercontinuum light sources include low functioning costs, high beam quality, and high efficiency in a maintenance-free format with a small footprint and low weight. Various bio-hazardous gases comprised of methane and hydrogen halides have their absorption in the near-infrared region. In the gas sensing process, gas is allowed to enter into the surrounding holes in the PCF as it will absorb the evanescent light from the core fiber. Depending on the intensity of the absorption wavelength, the output power is reduced strongly at the wavelength of interest. A similar mechanism is also followed for sensing fluids and biomolecules. Similarly, PCF-based laser and PCF-based sensor research is inspiring the sensor market, as within a few years it could share in a substantial way and are anticipated to be the industry standard.

There are numerous advantages to using OAM fibers in photonic devices for specific applications. The designing of OAM fiber is much more significant for the development of low-cost and unsophisticated photonic devices. Especially for obtaining the large number of OAM modes, recovering the initial signal, and transmitting signals over a long distance are still challenging. Introduction of OAM fiber in space division multiplexing (SDM) could offer a considerable improvement over merely using a ribbon of the fibers. Multiplexing various OAM channels permits high-capacity optical communication. By developing photonic integrated circuits (PIC) for OAM generation, multiplexing, and demultiplexing OAM signals, a single OAM fiber could replace a dozen single-mode fibers and accompanying connectors, sinking both space and complexity of the photonic system [2]. This could be employed inside a computer, using very short fiber (<1 m) but supporting numerous OAM modes; within a reconfigurable switch that connects various fiber ports; to link the components in the supercomputer, using short fibers (<20 m); or to link switches of a metropolitan access network (MAN), using short to medium-length fibers (<1 km). In the future, we will have lots of state-of-the-art photonic devices using OAM fibers. An optical device with various fiber ports can be substituted by a smaller device with a single OAM port. Further, a number of single-mode fiber with amplifiers can be replaced by a single OAM fiber link with a single amplifier per fiber span. A single OAM fiber-based amplifier with a single pump could amplify all channels in an OAM fiber, in contrast to one fiber amplifier and one pump laser per fiber needed for a ribbon of single-mode fibers. Therefore, cost-saving photonic devices would be possible using OAM fiber-based amplifiers. In this way, the future of OAM fibers is strongly interrelated to the development of devices able to work with those fibers: multiplexers and demultiplexers, switching, and amplifier. Though, optical scattering from the ambient micro-particles in the atmosphere or the mode coupling in the optical fibers considerably reduces the orthogonality between OAM channels for demultiplexing and ultimately increases the crosstalk in an optical communication system. This problem can be solved by implementing the scattering-matrix-assisted retrieval technique (SMART) to demultiplex OAM channels from highly scattered optical fields and achieve experimental crosstalk of −13.8 dB in the parallel sorting of 24 OAM channels after passing through a scattering medium [3]. To encode OAM channels the SMART can be applied in a self-built data transmission system that uses a digital micromirror device. Further, the SMART can also be applied to realize reference-free calibration. A high-fidelity transmission of both the colour and grey images under scattering conditions at an error rate of <0.08% has been reported already [3]. It is expected to accomplish high-performance optical communication in turbulent environments through this technique in the future.

Furthermore, the coreless PCF continuously twisted along its length, is able to induce robustly guide light forming a helical structure [4–6]. Since the effective mode area of the propagating mode shrinks and the effective refractive index rises with increasing the twist rate, it would be possible to develop a PCF design whose mode field diameter changes radically with propagation distance. This effect would be of interest in the delivery of high-power laser light. Throughout the transmission, the mode area could be kept large by applying a low twist rate and then focused to a smaller mode area near the end by increasing the twist rate. This effect would be of curiosity in the sensing applications. At the sensing location, the twist rate can be close to zero by simply reducing the twist rate to permit the modal field to spread out to the edge of the cladding. As we are aware, it is very difficult to achieve a large mode area with anomalous dispersion characteristics in conventional optical fibers. However, the fascinating feature of the twisted optical fiber is the combination of a large mode area with anomalous dispersion. This fascinating feature suggests the applications of twisted optical fibers in nonlinear optics. It is conventionally possible to have the fundamental solitons with very high peak powers using properly designed twisted optical fiber. The capability of the twisted optical fibers to offer circular birefringence and OAM advises that yet additional opportunities may appear from this exceptional and unanticipated guidance mechanism.

The availability of the air channels along the entire length of PCF has also unlocked the opportunity for functionalization of the channel surfaces at the molecular and the nanometre scales, in particular, to impart the functionality of surface-enhanced Raman scattering (SERS) in PCF for sensing and detection. PCF structure can work as an opto-fluidics platform, which makes it an excellent prospect for various technological and scientific applications. The SERS sensor based on PCF can be developed based on metal nanoparticle colloids. The metal nanoparticles and the analyte solution can be injected into the air holes of the PCF structure using a simply modified syringe to overcome mass-transport constraints [7, 8]. This method allows more metal nanoparticles involved in SERS activity. Such PCF-based sensor offers a noteworthy benefit over the conventional SERS sensor with high flexibility and easy manufacturing.

The microparticles can be guided optically along the hollow-core optical PCFs. The flying-particle sensors based on hollow-core PCFs can be a new paradigm in the reconfigurable optical fiber sensor for high accuracy multi-functional mapping over long distances [9]. By employing the parametric feedback or by spinning microparticles in the circularly polarized laser beam, the movement of the trapped microparticle could also be stabilized [10, 11]. The PCF sensing system exclusively permits multi-parameter sensing with high resolution in a single length of the optical fiber. Such hollow-core PCF-based sensors are of particular interest for use in highly radioactive environments, such as inside a nuclear reactor, where conventional solid-glass cores quickly darken because of the radiation damage.

The integration of the metamaterials on optical fiber allows wave guidance through the subwavelength geometry. Anisotropic metamaterial optical fiber offers a unique characteristic in which the modal fields can be controlled to confine near the centre of the waveguide due to the refractive index inhomogeneity [12]. Metamaterial-loaded waveguides for microwaves can support propagation at frequencies well below the cutoff. The inhomogeneity and easy control over the dielectric anisotropy made feasible in the nanoporous alumina fiber makes it an interesting candidate for the nano-photonic applications. The absorption loss and the instability of noble metal can be mitigated using circularly shaped metamaterial optical fiber with the core as crystalline germanium [13]. The metamaterial-based photonic crystal fiber can be used as an optical memory useful for the future all-optical computer [14]. The novel photonic technology of metasurfaces is able to open the door to a variety of optical fiber-based inventions. Ultrathin optical metalenses can be patterned directly on the facet of PCF that allows light focusing in the telecommunication regime [15, 16]. The ultrathin PCF-based metalenses may find various potential applications in different fields including optical sensing, imaging, and fiber laser designs.

REFERENCES

[1] Y. Q. Zong, S. W. Brown, B. C. Johnson, K. R. Lykke, and Y. Ohno, "Simple spectral stray light correction method for array spectroradiometers," *Appl. Opt.* 45, 1111 (2006).

[2] N. Zhang, K. Cicek, J. Zhu, S. Li, H. Li, M. Sorel, X. Cai, and S. Yu, "Manipulating optical vortices using integrated photonics," *Front. Optoelectron.* 9(2), 194–205 (2016).

[3] L. Gong, Q. Zhao, H. Zhang, X.Y. Hu, K. Huang, J. M. Yang, and Y. M. Li, "Optical orbital-angular-momentum multiplexed data transmission under high scattering," *Light Sci. Appl.* 8, 27 (2019).

[4] X. M. Xi, T. Weiss, G. K. L. Wong, F. Biancalana, S. M. Barnett, M. J. Padgett, and P. S. J. Russell, "Optical activity in twisted solid-core photonic crystal fibers," *Phys. Rev. Lett.* 110, 143903 (2013).

[5] X. M. Xi, G. K. L. Wong, M. H. Frosz, F. Babic, G. Ahmed, X. Jiang, T. G. Euser, and P. S. J. Russell, "Orbital-angular-momentum-preserving helical Bloch modes in twisted photonic crystal fiber," *Optica* 1, 165–169 (2014).

[6] R. Beravat, G. K. L. Wong, M. H. Frosz, X. Ming Xi, and P. S. J. Russell, "Twist-induced guidance in coreless photonic crystal fiber: A helical channel for light," *Sci. Adv.* 2(11), e1601421 (2016).

[7] X. Z. Guo, L. Y. Hua, W. Pei, L. K. Qun, Y. Jie, and M. Hai, "Photonic crystal fiber SERS sensors based on silver nanoparticle colloid," *Chin. Phys. Lett.* 25(12), 4473 (2008).

[8] B. Y. Han, S. Tan, M. K. K. Oo, D. Pristinski, S. Sukhishvili, and H. Du, "Towards full-length accumulative surface-enhanced Raman scattering-active photonic crystal fibers," *Adv. Mater.* 22, 2647–2651 (2010).

[9] D. S. Bykov, O. A. Schmidt, T. G. Euser, and P. S. J. Russell, "Flying particle sensors in hollow-core photonic crystal fiber," *Nat. Photon.* 9, 461 (2015).

[10] J. Gieseler, B. Deutsch, R. Quidant, and L. S. Novotny, "Parametric feedback cooling of a laser-trapped nanoparticle," *Phys. Rev. Lett.* 109, 103603 (2012).

[11] Y. Arita, M. Mazilu, and K. Dholakia, "Laser-induced rotation and cooling of a trapped microgyroscope in vacuum," *Nature Commun.* 4, 2374 (2013).

[12] D. Pratap, A. Bhardwaj, and S. A. Ramakrishna, "Inhomogeneously filled, cylindrically anisotropic metamaterial optical fiber," *J. Nanophoton.* 12(3), 033002 (2018).

[13] P. Mahalakshmi, S. A. Prakash, and M. S. M. Rajan, "Design of germanium core with anisotropic metamaterial cladding optical fiber in mid-infrared range applications," *Opt. Quantum Electron.* 52, 298 (2020).

[14] G. Palai, B. Nayak, S. K. Sahoo, S. R. Nayak, and S. K. Tripathy, "Metamaterial based photonic crystal fiber memory for optical computer," *Optik* 171, 393–396 (2018).

[15] J. Yang, I. Ghimire, P. C. Wu, S. Gurung, C. Arndt, D. P. Tsai, and H. W. H. Lee, "Photonic crystal fiber metalens," *Nanophotonics* 8(3), 443–449 (2019).

[16] M. Kim and S. Kim, "High efficiency dielectric photonic crystal fiber metalens," *Sci. Rep.* 10, 20898 (2020).

Index

Pages in *italics* refer to figures and pages in **bold** refer to tables.

Printed in the United States
by Baker & Taylor Publisher Services